Antimicrobials in Agriculture

This book offers comprehensive coverage of all manifestations of resistance in combating infectious diseases and explores advances in antimicrobials in agriculture and their applications in the fight against microbes. According to the World Health Organization, antimicrobial resistance is a major threat to global health because the number of alternative antibiotics is very limited. Antimicrobial resistance is a slow, evolutionary process that has been accelerated by human activities in the health, environment, and agriculture sectors. Due to their wide application, antibiotics and their residues have been found in almost all food products and natural ecosystems. This book appraises the drivers, impact, and mitigation of antimicrobials, with a focus on methods and targets. In addition, it also provides a variety of photographs, diagrams, and tables to help illustrate the material. The novel strategies to combat antimicrobial resistance are also described, emphasizing collaborative measures of control. The underlying molecular mechanisms, which depend not only on the microbe but on the specific drug molecule, are highly diverse and are covered in detail.

Students, researchers, scientists, practitioners, academicians, biologists, microbiologists, stakeholders, and policymakers can benefit from current trends in antimicrobials in agriculture that address microbiology, microbial biotechnology, ethnopharmacology, toxicology, natural medicinal plant products, secondary metabolites, and all disciplines related to antimicrobial research.

Features of the book:

- Covers antimicrobials in agriculture with up-to-date research.
- Recent references on plausible antimicrobials in agriculture.
- Public health impact of the use of antibiotics in agriculture.
- Antimicrobial efficacy of medicinal plants.
- Role of phytoalexins in agriculture.
- Nanoparticles as antimicrobial agents.
- Presents cutting-edge research on microbiology, nanotechnology, and emergent antimicrobial technologies.

Current Trends in Antimicrobial Research

Series Editor: Arti Gupta

Antimicrobials in Pharmaceutical and Medicinal Research
Edited by Arti Gupta and Ram Prasad

Antimicrobials in Food Science and Technology
Edited by Arti Gupta and Ram Prasad

Antimicrobial Resistance in Agriculture and its Consequences
Edited by Arti Gupta and Ram Prasad

Antimicrobials in Agriculture
Edited by Arti Gupta and Ram Prasad

For more info: https://www.routledge.com/Current-Trends-in-Antimicrobial-Research/book-series/CTAR

Antimicrobials in Agriculture

Edited by
Arti Gupta and Ram Prasad

CRC Press
Taylor & Francis Group
Boca Raton London New York

CRC Press is an imprint of the
Taylor & Francis Group, an **informa** business

Designed cover image: Shutterstock

First edition published 2024
by CRC Press
2385 NW Executive Center Drive, Suite 320, Boca Raton FL 33431

and by CRC Press
4 Park Square, Milton Park, Abingdon, Oxon, OX14 4RN

CRC Press is an imprint of Taylor & Francis Group, LLC

ISBN: 978-1-032-21542-6 (hbk)
ISBN: 978-1-032-21543-3 (pbk)
ISBN: 978-1-003-26889-5 (ebk)

DOI: 10.1201/9781003268895

Typeset in Times
by Newgen Publishing UK

Contents

Preface

Antimicrobials have been used in agriculture for decades, with recent estimates revealing that 73 per cent of all antibiotics worldwide are used in agriculture. Agricultural dependence on antimicrobials suggests habitual communication patterns, which are socially and culturally ingrained. Moreover, several reports indicate that antimicrobials are used imprudently in agriculture. For example, antimicrobials are frequently misused for growth advancement and preventative purposes. Antimicrobial resistance in agriculture is a valuable industrial resource that addresses multifaceted topics regarding farm, wild, friendly animals, and how the environment plays an active role in the amplification and transmission of resistant genes in the human food chain. This book explores the various aspects of microbials in agriculture and their perceived impact on human health including updated information about all the genera of livestock and associated pathogenic microbes. The valuable information may be exploited by policymakers and various groups of experts including academic and research personnel engaged in clinical and medical microbiology, along with all veterinary and medical practitioners. Researchers, scientists, and healthcare and management practitioners will find this an essential resource on the legal use of antibiotics in animals and humans.

The first chapter, by Phuengjayaem et al., reviews the antimicrobial resistance in lactic acid bacteria. In Chapter 2, Jain and Dey describe the antimicrobial indexing of indigenous pathogens from sanitary landfill leachate. Extensive studies and application of phytoalexins in agriculture discussed by Sheikh et al. in Chapter 3 and in Chapter 4, Kumar and his collaborators highlight that livestock manure application causes the spread of antibiotic-resistant genes in agricultural lands. In Chapter 5, Shedge et al. emphasize antibiotic residues in freshwater: antimicrobial resistance development and bioremediation. In Chapter 6, Jain and Pundir describe production of antimicrobial substances from endophytic microorganisms and potential applications as biocontrol agents, and in Chapter 7, Singh and Singh highlight the antimicrobial efficacy of medicinal plants in reference to present status and future research. In Chapter 8, Tajane et al. highlight the nano-materials as alternative antimicrobial agents in the management of phytopathogens in agricultural systems. Finally, nanotechnology-based antimicrobials in agriculture are more prominently explained in Chapter 9 by Thenmozhi Manivel et al.

Arti Gupta and Ram Prasad

Contributors

Anam Aslam
Department of Biochemistry, University of Lucknow, Lucknow, India

Thirumalaiyammal B.
Research Scholar PG and Research Department of Microbiology, Cauvery College for Women (Autonomous), Thiruchirappalli, Tamilnadu, India

Pritam Dey
CSIR-National Environmental Engineering Research Institute (NEERI), Kolkata Zonal Centre, i-8, sector-C, EKDP, E.M. Bypass, Kolkata, India

Ranjana Gautam
Department of Biochemistry, University of Lucknow, Lucknow, India

Ankita Gupta
Department of Biochemistry, University of Lucknow, Lucknow, India

Shefali Gupta
Department of Industrial Microbiology, Sam Higginbottom University of Agriculture, Technology and Sciences, Naini, Prayagraj, Uttar Pradesh, India

Pranay Jain
Department of Biotechnology, University Institute of Engineering and Technology, Kurukshetra University, Kurukshetra, Haryana, India

Rachna Jain
CSIR-National Environmental Engineering Research Institute (NEERI), Kolkata Zonal Centre, i-8, sector-C, EKDP, E.M. Bypass, Kolkata, India

Jayalaxmi
Department of Botany, St Aloysius College (Autonomous) Mangaluru, India

Dileep Kumar
Department of Biotechnology, Khwaja Moinuddin Chishti Language University, Lucknow, India

Ram Kumar Pundir
Department of Biotechnology, University Institute of Biotechnology, Chandigarh University, Mohali, Punjab, India

Swaroop Kumar V. S.
Department of Botany, St Aloysius College (Autonomous) Mangaluru, India

Hariprasad Paikrao
Government Institute of Forensic Science, Nagpur, RT Road, Civil Lines Nagpur, Maharashtra, India

Nidhi V. Maheshwari
Department of Biotechnology, S K Somaiya College, Somaiya Vidyavihar University, Vidyavihar, Mumbai, India

Jenjuiree Mahittikon
Department of Biochemistry and Microbiology, Faculty of Pharmaceutical Sciences, Chulalongkorn University, Bangkok, Thailand

Thenmozhi Manivel
Department of Microbiology, Indra Ganesan College of Arts And Science, Thiruchirappalli, Tamilnadu, India

Jyothi Miranda
Department of Botany, St Aloysius College (Autonomous) Mangaluru, India

Vishal Moger
Department of Botany, St Aloysius College (Autonomous) Mangaluru, India

Steffi P. F.
PG and Research Department of Microbiology, Cauvery College for Women (Autonomous), Thiruchirappalli, Tamilnadu, India

Anita Patil
Sant Gadge Baba Amravati University, Mardi Raod, Amravati. Maharashtra, India

Sukanya Phuengjayaem
Department of Microbiology, Faculty of Science, King Mongkut's University of Technology Thonburi, Bangkok, Thailand

Seema Sambrani
Department of Biotechnology, S K Somaiya College, Somaiya Vidyavihar University, Vidyavihar, Mumbai, India

Sathammaipriya
PG and Research Department of Microbiology, Cauvery College for Women (Autonomous), Thiruchirappalli, Tamilnadu, India

Vaishnavi Shedge
Department of Biotechnology, S K Somaiya College, Somaiya Vidyavihar University, Vidyavihar, Mumbai, India

Sana Sheikh
Department of Botany, St Aloysius College (Autonomous) Mangaluru, India

Sareen Sheikh
Department of Chemistry, Yenepoya Institute of Arts, Science and Management, Yenepoya deemed to be University, Mangaluru, India

Aniket Singh
Department of Zoology, Syamsundar College, Purba Bardhaman, West Bengal, India

Neela Singh
Department of Botany, Vinoba Bhave University, Hazaribag, Jharkhand, India

Diksha Tajane
Mustard Agronomist, AICRP on Linseed and Mustard, Dr. Punjabrao Deshmukh Krishi Vidyapeeth Akola, MS, India

Somboon Tanasupawat
Department of Biochemistry and Microbiology, Faculty of Pharmaceutical Sciences, Chulalongkorn University, Bangkok, Thailand

Anurag Yadav
Department of Microbiology, College of Basic Science and Humanities, S. D. Agricultural University, Banaskantha, Gujarat, India

Kusum Yadav
Department of Biochemistry, University of Lucknow, Lucknow, India

Editor Biography

Arti Gupta, PhD, obtained her PhD from Mahatma Jyotiba Phule Rohilkhand University, Bareilly, India, in 2010 in Animal Science. Dr. Gupta's current research interests are animal biotechnology, molecular plant biotechnology, molecular animal biotechnology, bioprocess technology, and microbiology. From July 2014 to till date, Dr. Gupta has been serving at Shri Avadh Raj Singh Smarak Degree College, Gonda, Uttar Pradesh, India, and has been engaged in some editorial work with Springer Nature, CRC Press and De Gruyter. Dr. Gupta has 12 years of teaching and research experience in plant and animal biotechnology, and microbial biotechnology. Dr. Gupta has worked as an external examiner for central evaluation in 2023–24 at MJP Rohilkhand University, Bareilly.

Dr. Gupta has received the following awards: University Topper (Gold Medal), MSc (Biotechnology) by Ch.C.S. University, Meerut; Young Scientist Award (Gold Medal) by the Zoological Society of India, Lucknow; Best Poster Presenter by Asian Journal of Experimental Science, Jaipur; Best Poster Presenter by International Consortium of Contemporary Biologists (ICCB) and Madhawai-Shyam Educational Trust, Ranchi; Fellowship Award by International Consortium of Contemporary Biologist (FICCB) and Madhawi Shayam Educational Trust (FMSET); and Dr. V.P. Agarwal (Gold Medal) by D.A.V. (P.G.) College, Muzaffarnagar. She has been involved with the following publications: 2 monographs, 12 edited books with Springer Nature, Switzerland, CRC Press T&Y USA and GmBH Germany; 7 book chapters, and around 50 national and international research papers. She has presented 28 abstracts in national and international symposia/seminars/conference/ workshops. Dr. Gupta has lifetime membership of the Indian Science Congress Association, Biotech Research Society of India, Zoological Society of India, and International Consortium of Contemporary Biologist. She has been serving as a member on several editorial boards. Previously, Dr. Gupta served as a member of teaching personnel at G.B. Pantnagar University, Pantnagar, Uttarakhand, India, and a visiting academic researcher at Biorefining and Advanced Material Research Center (BAMRC), and Scotland's Rural College (SRUC), University of Edinburgh, Scotland, United Kingdom

Ram Prasad, PhD is associated with the Department of Botany, Mahatma Gandhi Central University, Motihari, Bihar, India. His research interests include applied and environmental microbiology, plant-microbe – interactions, sustainable agriculture, and nanobiotechnology. Dr. Prasad has more than 300 publications to his credit, including research papers, review articles, book chapters, and 7 patents issued or pending, and has edited or authored several books. He has guided one postdoc, two M. Phil, nine Ph.D. and several master dissertation students. Dr. Prasad has fourteen years of teaching/research experience and has been awarded the Sir CV Raman Scientist Award, Young Scientist Award & Prof. J.S. Datta Munshi Gold Medal by the International Society for Ecological Communications; Fellow of Agricultural Technology Development Society (FATDS); Fellow of Biotechnology Research society of India

(FBRS); Fellow of the Society for Applied Biotechnology (FSAB); the American Cancer Society UICC International Fellowship for Beginning Investigators, USA; Outstanding Scientist Award in the field of Microbiology; BRICPL Science Investigator Award, Research Excellence Award etc. He has been serving as a member of several editorial boards. Previously, Dr. Prasad served as assistant professor at Amity University Uttar Pradesh, India; visiting assistant professor at Whiting School of Engineering, Department of Mechanical Engineering at Johns Hopkins University, Baltimore, United States, and research associate professor at the School of Environmental Science and Engineering, Sun Yat-sen University, Guangzhou, China.

Acknowledgements

First and foremost, we would like to thank GOD for his never-ending grace, mercy, and provision during what ended up being one of the toughest times of our life.

We are indebted to the many people who helped to bring this book to light. I wish to thank series editors Dr. Gustavo Molina and Dr. Vijai Kumar Gupta; Dr. Marc Gutierrez and Dr. Bonita Glanville-Morris, senior production editor Dr. Jyotsna Jangra and commissioning editor Dr. Renu Upadhyaya, and Taylor & Francis/CRC Press for their generous assistance, constant support, and patience in initializing the series volume. The editors are very thankful to the vice chancellors of Faizabad University, Ayodhya and Mahatma Gandhi Central University, Motihari, for their kind support and constant encouragement. Special thanks are due to my well-wishers, family members, and friends.

Dr. Arti Gupta gives special thanks to her exquisite husband, Mr. Saket Agrawal, and her son, Reyansh, who encouraged her to start the work, preserve, and put everything together. Dr. Ram Prasad gives his special regards to his family for their constant invaluable support, and motivation to publish it. We both also give special thanks to all faculty colleagues and senior faculties of our team.

Arti Gupta, *Shri Avadh Raj Singh Smarak Degree College, Gonda, UP, India*
Ram Prasad, *Mahatma Gandhi Central University, Motihari, Bihar, India*

1 Antimicrobial Resistance in Lactic Acid Bacteria

Sukanya Phuengjayaem, Jenjuiree Mahittikon, and Somboon Tanasupawat

1.1 INTRODUCTION

Currently, antimicrobials are widely used in household products, agriculture, and medicine. Excessive usage of antimicrobial agents is focused on cost-saving and low toxicity of medication selection in drugs (Griffith et al., 2012). Most of the antimicrobials used in medicine are derived from secondary metabolites of plants, fungi, bacteria, and marine organisms. Microbes can exist in a mutually beneficial relationship. Still, when nutrients are in short supply, they can form an antagonistic connection, resulting in the generation of harmful chemicals and metabolites for other species (Bode, 2009). Natural products have been used as a utilized medicine to treat a variety of diseases such as cancer, hypercholesterolemia, and immune dysfunction (Wang et al., 2018; Gora et al., 2021; Huang, Lu and Ding, 2021).

Antimicrobial usage worldwide between 2000 and 2015, antibiotic consumption climbed by 65%, while the antibiotic use rate increased by 39% (Klein et al., 2018). Furthermore, global antimicrobial usage is expected to rise by 67% from 2010 to 2030, from approximately 63 thousand tons to over 100 thousand tons (Van Boeckel et al., 2015). Changing production techniques in middle-income countries has contributed to a one-third increase in livestock consumption between 2010 and 2030. Large-scale intensive agricultural enterprises that habitually employ antimicrobials at subtherapeutic levels will replace vast farming systems (Van Boeckel et al., 2015). Even though it is arduous to evaluate the actual numbers of antimicrobial agents used because of the distinction in variability and international registration system, the continued increase global sale of antibiotics could indicate the trend of antibiotic accumulation in the environment (Tiseo et al., 2020). This information may be linked to a global assessment on antibiotic resistance, which predicts that by 2050, with a population of 10 million, there will be one fatality per year (Tong, 2021). Besides the utilization of antimicrobials, the dangers of antimicrobial resistance in microorganisms are the global faces. They are the most serious worldwide health issues. Uncontrolled use of antimicrobial reagents and disposal of wastes without waste categorization are causes of antimicrobial contamination and antimicrobial resistance distribution to the environment (Kimera et al., 2020). Therefore, the study of the involvement between antibiotic usage, the generation, the dispersal of antimicrobial resistance, and multi-resistant bacteria should be considered (Busetti et al., 2017).

Lactic acid bacteria (LAB) are microorganisms widely employed in food manufacturing as probiotics. They are living organisms that provide health advantages to the host (Korhonen, 2010). Therefore, they probably account for the most natural gut microbiome of humans and animals (Álvarez-Cisneros and Ponce-Alquicira, 2018). Currently, the demand for probiotic product usage is a rising trend because most people are concerned about improving their quality of life (Colombo et al., 2018). Good nutrition, physical activity, and a healthy lifestyle are connected to good health and well-being (Ouwehand and Röytiö, 2015). *Lactobacillus* (*Lb*), *Lactococcus* (*Lc*), *Pediococcus* (*P*), *Streptococcus* (*S*), *Enterococcus* (*E*), and *Bifidobacterium* (*B*) strains are employed as probiotics. They are usually found in foods and dietary supplements. *Lactobacillus bulgaricus*

DOI: 10.1201/9781003268895-1

and *Streptococcus thermophilus* strains are commonly used in yogurt production (Adolfsson et al., 2004). *Lactobacillus*, *Lactococcus*, *Leuconostoc*, and *Pediococcus* strains are regularly used in dairy products, for example, milk preservation, cheese, kefir, and buttermilk (Clementi and Aquilanti, 2011; Rakhmanova et al., 2018). Additionally, LAB can be used to deliver vaccines and other substances directly into the digestive tract system (Gad et al., 2014).

Although antibiotic-resistant properties of LAB are useful to humans or animals by assisting in maintaining gastrointestinal equilibrium in situations of antibiotic-induced diarrhoea, they may serve as a reservoir of antibiotic resistance genes that could be transferred onto pathogenic bacteria (Erginkaya et al., 2018). LAB antimicrobial resistance in food products has been discovered as either intrinsic or acquired antimicrobial agents (Mathur and Singh, 2005).

Many antibiotics are thought to be inherently resistant to LAB (Preethi et al., 2017). LAB could be resistant to other antimicrobial agents, thereby spreading resistance to diseases found in the gastrointestinal tracts of both animals and humans. Antibiotic-resistant food-borne organisms have largely been reported, for example, antimicrobial resistance in LAB, isolated from Chinese yogurts (Zhou et al., 2017), traditional Turkish fermented dairy (Erginkaya et al., 2018), faeces of broiler chicken in Malaysia (Shazali et al., 2014). In addition, antibiotic resistance in *Lactobacillus*, *Pediococcus*, and *Enterococcus* strains could be isolated from dry-fermented sausages (Fraqueza, 2015). They contributed to the growing wealth of evidence demonstrating the major function of food-borne LAB as a horizontal reservoir of antibiotic-resistant genes that are transferred horizontally through the food chain to pathogens, posing a significant public health concern. Mostly antimicrobial-resistant, vancomycin has been observed in *Enterococcus*, *Lactobacillus*, *Lactococcus*, *Pediococcus*, and *Bifidobacterium* species (Temmerman et al., 2003; Jia et al., 2014; Rozman et al., 2020). Furthermore, bacitracin, cefoxitin, ciprofloxacin, fusidic acid, streptomycin, sulphadiazine, teicoplanin, and vancomycin resistance is intense in some lactobacilli (Danielsen and Wind, 2003).

This raises concerns that the biggest threat presented by antibiotic-resistance genes could be transmitted to humans via the food chain (Flórez et al., 2005; Rodriguez-Alonso et al., 2009). Especially, unheated fermented milk products serve as a medium for antibiotic-resistant bacteria. Indigenous bacteria and the human gastrointestinal tract have a direct relationship (Mathur and Singh, 2005). Cheese made from raw milk contains antimicrobial-resistant LAB, which may act as a vector to create resistance to certain antibiotics in bacteria humans infect (Erginkaya et al., 2018). As a result, the increased use of LAB in food manufacture as probiotics during the previous few decades has raised safety risks worldwide (Ammor et al., 2007). The significant threat to antibiotic usage is establishing and spreading resistance in pathogenic bacteria that have previously failed to respond to previously effective treatments (Erginkaya et al., 2018).

Most of the research has concentrated on historically clinically relevant bacterial strains. However, there has been an increase in interest in isolated strains that have not been clinically proven in recent years, particularly LAB, which accumulate several antibiotic-resistant genes. Because LAB are generally used in the food industry, they may transfer that antibiotic-resistant gene to pathogenic bacteria and humans and animals (Rodriguez-Alonso et al., 2009; Erginkaya et al., 2018). They could make it more difficult to treat a patient with an infection or disease resistant to antibiotics (Korhonen, 2010). To acquire a perspective of the current situation in LAB regarding antibiotic resistance trends, a wide range of beneficial bacterial strains belonging to *Bifidobacterium*, *Lactobacillus*, *Lactococcus*, *Enterococcus*, and *Pediococcus* species are summarized in this chapter.

1.2 ANTIMICROBIALS AND THEIR ACTION IN BACTERIAL CELLS

The antimicrobial categories include resisting cell wall synthesis, destroying cytoplasmic membranes, inhibiting protein synthesis, inhibiting nucleic acid replication, and interfering metabolic pathways of bacteria metabolism (Hooper, 2001; Reygaert, 2018; Álvarez-Cisneros and Ponce-Alquicira, 2018). The resistance cell wall synthesis is subdivided into three functions: preventing the polymerization and cross-linking process of cell wall peptidoglycan, blocking cell wall precursors from being transported through the cell membrane, and inhibiting peptidoglycan precursors. The case of destroying the cytoplasmic membrane is subdivided into three functions: depolarization of the cytoplasmic membrane, increasing membrane permeability, removing the membrane, and creating pores (Amerikova et al., 2019). Examples of medications and functions from each of these categories are described in Figure 1.1.

The mode of action of antimicrobials for clinical usage is divided into five groups according to the mechanism of antimicrobial agents, which is shown in Figure 1.1.

Although the various mechanisms are described above for improved management of pathogens, ineffective antimicrobial treatment has contributed to the numerous antimicrobial resistance problems that are now encountered. Excessive usage of antimicrobial agents since medication selection in drugs is focused on cost-saving and low toxicity (Griffith et al., 2012). Increased antimicrobial medication consumption, both by humans and animals, and unsuitably antimicrobial treatment prescribing are important factors contributing to the rising resistance problem (Victor, 2011).

The research of Morar et al. (2015) reported that microbiological and diagnostic evidence seems to indicate that angiotensin receptor blockers (ARBs) and antimicrobial resistance genes (ARGs) are frequently transferred from animals to humans who consume food animals. Ultimately, people are increasingly struggling with more severe diseases that require more extensive treatment and prolonged disease periods that frequently necessitate protracted hospitalization (Amerikova et al., 2019). The expenses of treating these illnesses have risen because of this. Currently, LAB is widely used in the fermented food industry (Amerikova et al., 2019). However, it is important to note that antimicrobial resistance genes are also discovered in the LAB. Therefore, it is extremely important

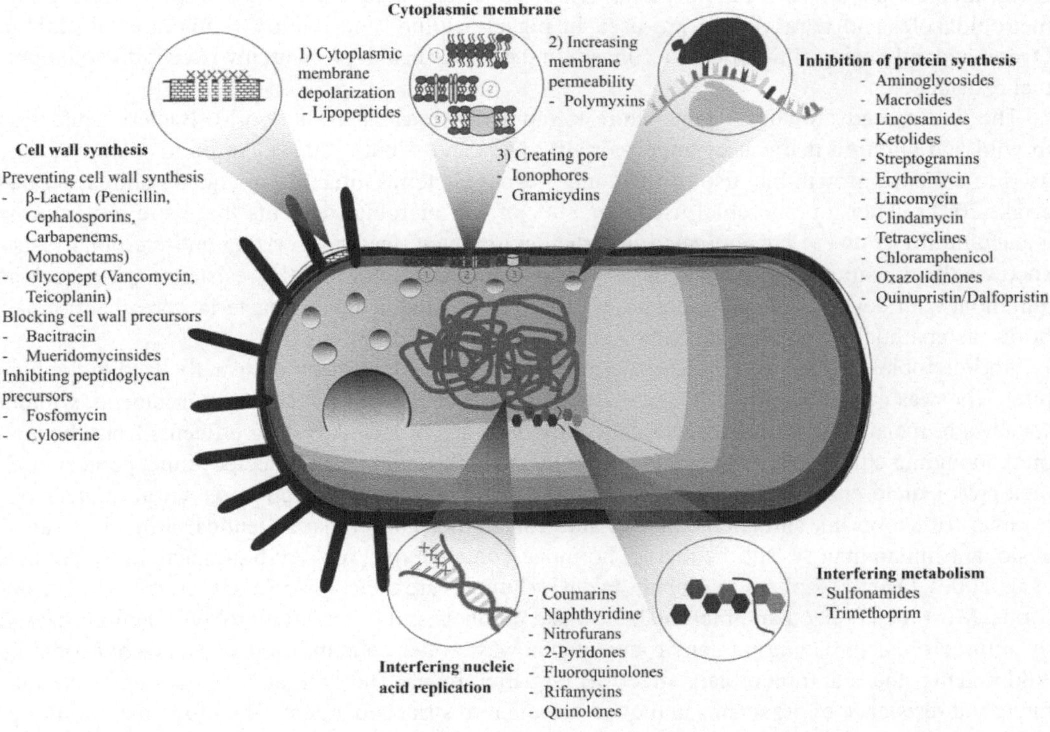

FIGURE 1.1 Antimicrobial action in bacterial cells.

Source: (Hooper, 2001; Reygaert, 2018; Álvarez-Cisneros and Ponce-Alquicira, 2018).

to pay attention to the dosage, waste disposal methods to the environment, and applications from bacteria that may have drug-resistance genes attached to them. The WHO recommends that LAB used in the food industry be free of resistance (Álvarez-Cisneros and Ponce-Alquicira, 2018).

1.3 SOURCE OF ANTIMICROBIAL CONTAMINATION AND ACCUMULATION IN THE ENVIRONMENT

1.3.1 Antimicrobial Contamination

Antimicrobials are discarded from the household via the trash bin, clothing, toilets, pharmacies, etc., when unused, expired, or when treatment is finished (Bound and Voulvoulis, 2005). The households release wastewater into suburban canals, city canals, and rivers, respectively. Antimicrobial lists are found in the aquatic environment comprised of sulfamethoxazole, sulfamethazine, sulfapyridine, trimethoprim, erythromycin, clarithromycin, roxithromycin, oleandomycin, oxytetracycline, tetracycline, and lincomycin (Binh et al., 2018). Furthermore, several personal care products containing triclosan, such as handwash soaps, act as a powerful antimicrobial selection source in residential wastewater (Jutkina et al., 2018).

Antibiotics have gradually increased use in the agricultural industries to treat animal infections caused by bacteria as therapeutic usage, to prevent bacterial infections as prophylactic usage, and to reduce management costs by decreasing investment (Kimera et al., 2020). Tetracyclines are used as growth promotors in livestock food additive forms (Rugumisa et al., 2016). Recently, sources of antimicrobial substances have been easily accessed from many resources such as official and unofficial dealers, who provide minimal information to farmers that initiate drug misuse. Human medicines are usually taken for therapeutic and prophylactic treatments. Human drugs (amoxicillin, ampicloxacillin, chloramphenicol, doxycycline, chloramphenicol eye drops, ciprofloxacin, metronidazole, and tetracycline) are used in pig agriculture and poultry (Kimera et al., 2020). Overabundantly using antibiotics is a concern in the public health problem involved with antimicrobial resistance.

The widespread growth of aquaculture is rapidly increasing in the world. Bacterial infection in wild and cultured fish causes huge mortality (Aly and Albutti, 2014). Antibiotics are not only used to enhance growth but also to treat and prevent bacterial infection in aquatic animals (Zhao et al., 2021). After antimicrobial use, over 80% of antimicrobial reagents that were taken in the aquacultural farm were contaminated into the environment, leading to rising antimicrobial resistance via the transition of resistance or multi-resistance genes. Further, the resistance genes such as quinolones, tetracyclines, and β-lactamases, which originated in aquatic bacteria, spread to normal-floras, fish pathogens, and human pathogens (Cabello et al., 2013).

Antimicrobial and pharmaceutical compounds for patient treatment are usually used in the hospital. The wastewater from the hospital is frequently released into natural environments without wastewater management. Pharmaceutical compounds are collected as drug effluents from the hospital, including chlorotetracycline, ciprofloxacin, cyclophosphamide, demeclocycline, paracetamol, metoprolol, diclofenac, ibuprofen, 17β-estradiol, estriol, estrone, doxycycline, oxytetracycline, tetracycline, trimethoprim, sulfamethoxazole, and ifosfamide. These pharmaceutical drugs have analgesic, anti-inflammatory, antirheumatic, hormone contraceptive, and antibiotic activities (Thomas et al., 2007). Over 90% of antimicrobials taken by humans are metabolized and secreted in the faeces forms. Most metabolized antimicrobials in waste products still have antimicrobial functions caused by antimicrobial spreading into environments via wastewater contaminated with resistant bacteria. Additionally, these antimicrobials affect pH, aquatic organic and inorganic compounds, and antimicrobial resistance of organisms in the marine and terrestrial ecosystem. Therefore, the multidrug-resistant bacteria from the hospital sewage are isolated and demonstrate the expansion of drug and multidrug resistance into the environment (Serwecińska, 2020). Pharmaceutical industries release a large number of antimicrobial residues and other co-factors, comprising metal and surfactants,

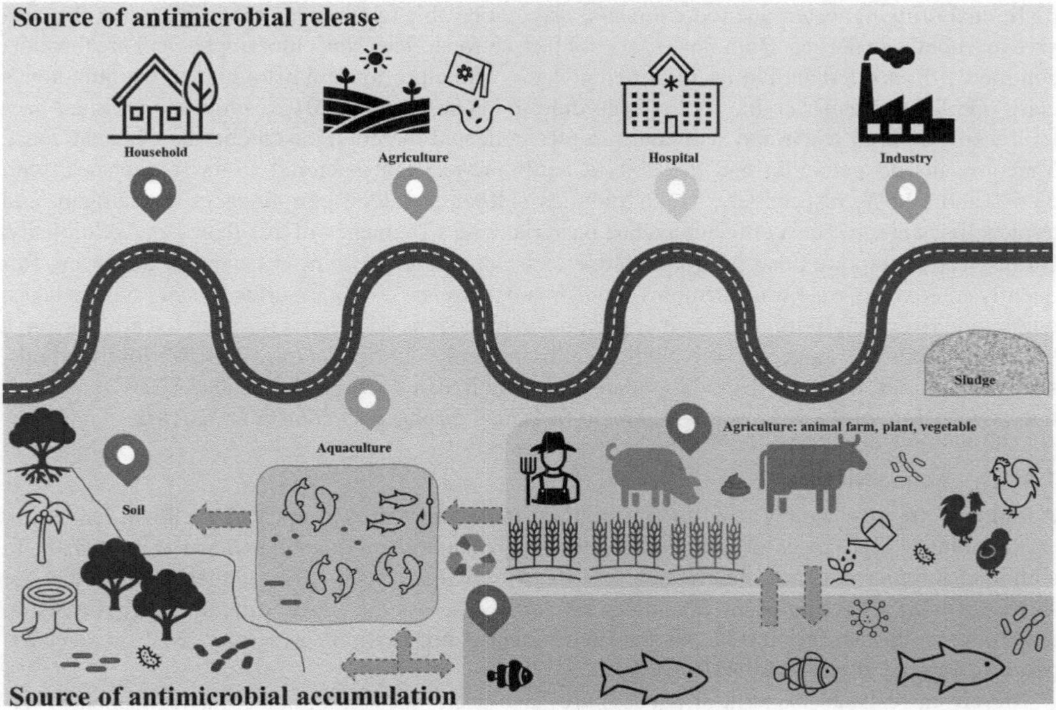

FIGURE 1.2 Source of antimicrobial release and accumulation in the environment.

into sewage (Pruden et al., 2013). Ofloxacin, sulfamethoxazole, trimethoprim, azithromycin, clarithromycin, cefotaxime, cephalexin, and cefixime from the pharmaceutical industry are reported to be released in the aquatic environment in Vietnam (Binh et al., 2018).

1.3.2 Antimicrobial Accumulation in the Environment

Eluents from wastewater treatment facilities, the pharmaceutical sector, the agricultural sector, and surface runoff from animal feeding activities are the main sources of antimicrobials in groundwater and surface water (Walters et al., 2010). Nevertheless, due to the high expenses of their usage, these procedures may not be used in all circumstances, and the amount of antimicrobial degradation varies depending on the category of pharmaceuticals and their compounds (Thiele-Brun and Peters, 2007; Mitchell et al., 2015). Furthermore, many traditional wastewater treatments in facilities are not removed and medications and antibiotic residues have also been discovered in the final leaching of wastewater treatment facilities throughout the world (Serwecińska, 2020). Especially hospital wastewater is the source of pathogens and antibiotics that have no normative status. In most countries, hospital wastewater is no separate category of wastewater that is particularly harmful to the environment and enters municipal wastewater treatment. As a result, for these pollutants, a monitoring system is required (Urbaniak et al., 2017). Sewage sludge is regularly distributed over the soil as a feature of agricultural operations because of its high number of organic compounds and nutrients. It is distinguished by high quantities of antibiotic resistance bacteria (ARB), and antibiotic resistance genes (ARG) expelled from entering wastewater. It might be harmful to the environment as well as human health (Calero-Cáceres et al., 2014; Koch et al., 2021).

Antibiotics have been usually found in surface water with concentrations in the range of 0.01 to 1.0 g/L. It indicates that they have migrated into a supply of drinking water sources and that traditional drinking water, due to treatments, may not be able to eliminate them. They can also survive for months in the sea. Quinolones, macrolides, sulfonamides, and chloramphenicol are the most commonly discovered antibiotics in drinking water, with ciprofloxacin belonging to the quinolones class (De Kwaadsteniet et al., 2006). According to Vignesh et al. (2016), antibiotic-resistant bacteria were found in water and sediment samples collected in ecoregions along the Chennai coast. Vancomycin and penicillin had the highest antibiotic-resistant bacterial strain frequencies, with 53.6% and 52.6%, respectively. The antibiotics chloramphenicol, ciprofloxacin, gentamicin, and tetracycline were the most efficient against bacteria, with a frequency of less than 7%. Additionally, Harnisz (2013) reported that fluoroquinolone-resistant and tetracycline-resistant bacteria were frequently entered into river water samples downstream (Harnisz, 2013). In surface waters such as lakes, ponds, and rivers, multidrug-resistant opportunistic pathogens with genes resistant to carbapenem, a last-resort antibiotic, pose a severe public health problem (Kittinger et al., 2018). Various methods, such as transport, dilution, degradation, and adsorption, can decrease ARGs and ARBs concentration downstream of the wastewater treatment facilities' discharges (Yoon et al., 2018).

1.3.2.1 Aquaculture

Antibiotics are also widely used in aquaculture, including fish, shrimp, and shellfish, leading to extraordinarily high levels of antibiotics entering the aquatic environment (Schar et al., 2020). In addition, antimicrobials are ineffective in fish. Approximately 70–80% of their metabolites are active antibiotics floating in fluids and can be diffused across the entire water system. As a result, antibiotic use in fish farming has major environmental implications, as it may result in selection pressure on several ecosystems (Burridge et al., 2010).

Moreover, resistant bacteria in aquaculture have been discovered, and it has expanded to terrestrial animals and humans. They can transfer their resistance genes to opportunistic infections in humans and animals (Serwecińska, 2020). Antimicrobial agents introduced in aquaculture are commonly used as metaphylaxis to whole populations consisting of individuals who are healthy, sick, or carriers, and in quantities that are usually higher than those used in land agriculture (Serwecińska, 2020). Hanna et al. (2018) reported that florfenicol, oxytetracycline, sulfadimethoxine, and ormetoprim have all been approved for aquaculture in the United States. In most European countries, erythromycin, florfenicol, oxytetracycline, ormetoprim, sarafloxacin, sulphonamides, and trimethoprim are permitted (Kümmerer, 2009). Until 2010, China had the largest manufacturing and use of antibiotics (Yang et al., 2010). The country was the top exporter of fish and fisheries products in the world (Serwecińska, 2020).

1.3.2.2 Soil

Bacterial species of clinical relevance are found in soils. Antibiotic pollution's selective pressure on antimicrobial growth in the soil is of specific importance (Tello et al., 2012). Sludge comprises xenobiotics and resistant bacteria that can affect the metabolic functions and the variety of the soil microbial community, lowering the quality of the soil ecosystem (Knapp et al., 2010). Antibiotic residues, both synthetical material and semi-synthetical material, including fluoroquinolones, tetracyclines, and sulfonamides, attach to soil particles. They have higher chemical stability and are resistant to bacterial digestion, which greatly leads to biodegradation delays. Therefore, they are frequently distributed in soil (Rusch et al., 2019). In addition, ciprofloxacin and oxolinic acid require five months for the removal of only 20% of oxolinic acid (Larsson et al., 2007). Recently, the investigation also necessitates the addition of ARB and ARG to the list of developing pollutants found in sediments and their elimination from sediments and fertilized soils (Wang et al., 2017).

1.3.2.3 Agriculture: Animal Farm, Plant, Vegetable

Antibiotics are used to treat disorders in animals, such as chronic enteritis, which leads to unhealthy food consumption and weight loss (Serwecińska, 2020). Mostly, antibiotics, chlortetracycline, flavophospholipid, oxytetracycline, penicillin, and bacitracin are widely used to prevent pathogen infections and treat animal diseases (Serwecińska, 2020).

Even though it is arduous to evaluate the actual numbers of antimicrobial agents used in animals that produce food because of the distinction in variability and international registration system, the global sale of medically essential antibiotics for use in food-producing animals reached 93,309 tonnes in 2017. It grew by 3% in 2019 (Tiseo et al., 2020). It is a critical problem since practically all active antimicrobial drugs used in animal agriculture are structurally similar to those used to treat humans, encouraging co-resistance and cross-resistance (EMA, 2014; ECDC et al., 2017).

Besides breeding animals (poultry, cattle, pigs, and rabbits), antibiotics are employed to eliminate pests in agricultural production, as well as beekeeping (McManus et al., 2002; Raymann et al., 2017). Streptomycin is authorized on peppers and tomatoes to control *Xanthomonas campestris* pv. *vesicatoria*. However, *Erwinia amylovora* is resistant to streptomycin (*SmR*), which produces fire blight and has economic and political ramifications (McManus et al., 2002). On the other hand, tetracycline (*tetA*, *tetB*, *tetC*, and *tetR*) is also used in animal husbandry and then released into the environment. Tetracycline-resistance genes (*TcR*) homologous to *tetA* and *tetM* have been discovered in a variety of plant-pathogenic bacteria (Sundin and Wang, 2018).

Furthermore, the WHO considers antibiotics such as penicillins, tetracyclines, aminoglycosides, macrolides, quinolones, and sulphonamides, which are extensively used in agriculture, to be particularly significant for human therapy (WHO, 2011). The research of Morar et al. (2015) reported that microbiological and diagnostic evidence seems to indicate that ARBs and ARGs are frequently passed down from animals that produce food for humans. Most of the research has been undertaken on the key parameters that drive the selection and transmission of antimicrobial resistance in food across the food chain.

Lactic acid bacteria (LAB) are utilized in several fields involved in our daily life. They are extensively used in food, chemical, pharmaceutical, and other industries (De Lima et al., 2010). On the contrary, antimicrobial resistance in microorganisms is one of the most serious worldwide health issues. Uncontrolled use of antimicrobial reagents and disposal of wastes without waste categorization from humans, households, hospitals, agriculture, aquaculture, industries, etc. (Figure 1.2), which are released into the environment, are causes of antimicrobial contamination and antimicrobial resistance distribution to the environments (Kimera et al., 2020). Many antibiotics are released into environments because their use enhances the development of a healthy modern society. The selective pressure of agriculture and animal care in the clinical setting increases concern about the human pathogen, which brings about research on the connection between antibiotic usage, the generation and dispersal of antimicrobial resistance, and multi-resistant bacteria (Busetti et al., 2017). Most antimicrobials released into the environment affect the resistance mechanism of lactic acid bacteria, including chloramphenicol, ciprofloxacin, erythromycin, gentamicin, vancomycin, rifampin, and tetracycline. Especially vancomycin resistance gene was found in LAB, including *Bifidobacterium*, *Lactobacillus*, *Enterococcus*, and *Streptococcus* strains (Erginkaya et al., 2018).

1.4 RESISTANCE OF ANTIMICROBIAL GENES, VIRULENT GENES, AND GENOMIC ANALYSIS IN LAB

1.4.1 *BIFIDOBACTERIUM*

Bifidobacteria are the probiotic members involved in the food chain. On the other hand, they are categorized as the antimicrobial resistance strain, one of the global public health concerns. Several genotypic characteristics of *Bifidobacterium* are observed for data collection and prediction of the antimicrobial resistance genes, mutation, translocation of genetic elements, etc. Flórez et al. (2006) examined antimicrobial susceptibility characteristics of *Bifidobacterium* strains screened by the minimal inhibitory concentration (MIC) method. The results show antimicrobial resistance phenotypes of this bacteria, including amikacin, ciprofloxacin (CIP), daptomycin (DAP), gentamicin (GEN), kanamycin (KAN), nalidixic acid (NAL), streptomycin (STR), tetracycline (TET) resistance. Nevertheless, the detection of the resistance gene of the *Bifidobacterium* strain via the PCR technique has shown only the TET resistance (*tetW*) gene from healthy human intestinal *B. longum* H66. Additionally, Ouoba et al. (2008) proposed the resistance gene of lactic acid bacteria, including bifidobacteria. They used the PCR method to amplify and sequence for resistance gene detection. The CIP resistance (*gyrA*) and STR resistance (*aadE*) genes represented the positive results in *B. longum* and *B. bifidum*. Interestingly, Jung et al. (2018) sequenced and read the whole-genome sequence of *B. choerinum* FMB-1, using HGAP 3.0, a hierarchical genome-assembly process DNA sequencing for high-quality *de novo* microbial genome assembly (Chin et al., 2013). This study identifies the antimicrobial resistance gene by MEGARes (Lakin et al., 2016). Three resistance genes BcFMB_02975, BcFMB_04310, and BcFMB_043, express as 3 S rRNA methyltransferases (preventing antibiotic activities), aminoglycoside O-nucleotidyltransferases (aminoglycosides resistant), lincosamide nucleotidyltransferases (lincosamide resistant; LIN), respectively. Esaiassen et al. (2017) analyzed phenotypic and genotypic characteristics from *Bifidobacterium* spp., including *B. longum* 11 strains, *B. breve* 2 strains, and *B. animalis* strains. They reported that one of *B. longum* resists LIN by expression of *lmrD* gene. Resistant of TET in *B. animalis* and *B. breve* cause by *tetT* gene. CIP resistance was found in *B. animalis* by *mfd*, *gyrA*, and *gyrB* gene expressions. Moreover, Rozman et al. (2020) characterized the antimicrobial resistance genes in bifidobacterial by using *in silico* detection, which combines several databases, bioinformatic instruments, etc. Antibiotic-resistant genes of *Bifidobacterium* strains are represented in Table 1.1.

Mucus display on pathogen adhesive prevention on the intestine. Ruas-Madiedo et al. (2008) reported two genes that involve intestinal mucin degradation. This study amplified two virulence genes from the *Bifidobacterium* strain by the PCR method as *engBF* and *afcA*: *engBF* of *B. longum* JCM1217 represent endo-α-N-acetylgalactosaminidase; *B. longum* JCM1254 encode *afcA* to produce 1,2-α-l-fucosidases. Esaiassen et al. (2017) expressed twenty virulence genes of bifidobacterial strains, comprised of iron and magnesium ion transportation, inducing bacterial adhesion, stress tolerance protein, immune-activated protein, and toxin secretion. In their results, *B. animalis* represents gene *iap*(*cwhA*), encoded into extracellular protein p60. Moreover, they investigate the virulence gene of *B. longum* subsp. *infantis*, which have the ability to encode *ureA* and *ureB* to produce urease alpha and beta subunits, respectively. These enzymes hydrolyze urea to produce ammonia and carbamate. In addition, they increase pH to survive and colonize the gastrointestinal tract.

1.4.2 *Lactobacillus*

Normally, antibiotics that impede protein synthesis are toxic to lactobacilli, including clindamycin (CLI), chloramphenicol (CHL), erythromycin (ERY), and tetracycline (TET) (Ammor et al., 2007; Klare et al., 2007). According to Bennedsen et al. (2011), the result from the *in silico* genome analysis was that *Lactobacillus* 8 species and 16 strains, including *Lb. delbrueckii* subsp. *bulgaricus* strain CHCC8942, CHCC769, CHCC5213; *Lb. fermentum* strain CHCC10568, CHCC5774, *Lb. paracasei* strain CHCC3136, CHCC2115, CHCC10665; *Lb. acidophilus* CHCC3777 *Lb. rhamnosus* CHCC5366, *Lb. rhamnosus* CHCC3402, CHCC5150, *Lb. reuteri* strain CHCC12039, and *Lb. plantarum* strain CHCC2365, CHCC10668, CHCC10672 had not detected antimicrobial resistance and virulence factor genes.

Similarly, most LAB are isolated from the environment of a bovine feedlot. The prevalence of susceptibility to ampicillin (AMP), CLI, CHL, GEN, KAN, and STR is greater than 80%. However, lactobacilli were detected to be moderately and highly resistant to ERY and TET at 48% and 79%, respectively (Aristimuño Ficoseco et al., 2018).

Additionally, the results of *Lactobacillus* sp. were isolated from chicken and boar droppings (Klose et al., 2014), including *Lb. rhamnosus*, *Lb. acidophilus*, *Lb. casei*, *Lb. plantarum*, and *Lb. mucosae*. They were all sensitive to the β-lactam AMP, and they were commonly utilized as a probiotic (Hummel et al., 2007; Klose et al., 2014). This result was consistent with Cauwerts

TABLE 1.1
Antibiotic Resistance Determinants in *Bifidobacterium* Species

Strain	Antibiotic	Representative gene	Genomic analysis	Reference
B. animalis subsp. *animalis* ATCC 25527	LIN	*lmrD*	Whole-genome sequencing and *in silico*	Rozman et al. (2020)
	VAN	*vanRM*		
	FO	*murA*		
	MU	*ileS*		
	FQ acridine dye	*arlR*		
	TET	*tetA(48)*		
	macrolide, LIN	*RlmA(II)*		
	peptide antibiotic	*ugd*		
	AMC	*parY*		
	macrolide, penam	*mtrA*		
	peptide antibiotic, rifamycin	*rpoB2*		
B. animalis subsp. *animalis* IM386	LIN	*lmrD*	Whole-genome sequencing and *in silico*	Rozman et al. (2020)
	VAN	*vanRM*		
	FQ	*murA*		
	mupirocin	*ileS*		
	FQ, acridine dye	*arlR*		
	TET	*tetA(48)*		
	macrolide, LIN	*RlmA(II)*		
	peptide antibiotic	*ugd*		
	AMC	*parY*		
	macrolide, penam	*mtrA*		
	peptide antibiotic, rifamycin	*rpoB2*		
B. animalis subsp. *lactis* Bi-07	LIN	*lmrD*	Whole-genome sequencing and *in silico*	Rozman et al. (2020)
	tiamulin	*tvaA*		
	macrolide	*macB*		
	VAN	*vanRM*		
	fosfomycin	*murA*		
	mupirocin	*ileS*		
	FQ, acridine dye	*arlR*		
	TET	*tetA(48)*		
	TET	*tetW*		
	macrolide, LIN	*RlmA(II)*		
	AMC	*parY*		
	macrolide, penam	*mtrA*		
	peptide antibiotic, rifamycin	*rpoB2*		
B. animalis subsp. *lactis* BB-12	tiamulin	*tvaA*	Whole-genome sequencing and *in silico*	Rozman et al. (2020)
	VAN	*vanRM*		
	fosfomycin	*murA*		
	mupirocin	*ileS*		
	FQ, acridine dye	*arlR*		
	TET	*tetA(48)*		
	macrolide, LIN	*RlmA(II)*		
	peptide antibiotic	*ugd*		
	peptide antibiotic	*ugd*		
	AMC	*parY*		
	macrolide, penam	*mtrA*		
	peptide antibiotic, rifamycin	*rpoB2*		

(continued)

TABLE 1.1 (Continued)
Antibiotic Resistance Determinants in *Bifidobacterium* Species

Strain	Antibiotic	Representative gene	Genomic analysis	Reference
B. animalis subsp. *lactis* Bl-04	tiamulin	*tvaA*	Whole-genome sequencing and *in silico*	Rozman et al. (2020)
	VAN	*vanRM*		
	fosfomycin	*murA*		
	mupirocin	*ileS*		
	FQ, acridine dye	*arlR*		
	TET	*tetA(48)*		
	TET	*tetW*		
	macrolide, LIN	*RlmA(II)*		
	peptide antibiotic	*ugd*		
	AMC	*parY*		
	macrolide, penam	*mtrA*		
	peptide antibiotic, rifamycin	*rpoB2*		
B. animalis subsp. *lactis* HN019	tiamulin	*tvaA*	Whole-genome sequencing and *in silico*	Rozman et al. (2020)
	VAN	*vanRM*		
	fosfomycin	*murA*		
	mupirocin	*ileS*		
	FQ, acridine dye	*arlR*		
	TET	*tetA(48)*		
	TET	*tetW*		
	macrolide, LIN	*RlmA(II)*		
	peptide antibiotic	*ugd*		
	AMC	*parY*		
	macrolide, penam	*mtrA*		
	peptide antibiotic, rifamycin	*rpoB2*		
B. choerinum FMB-1	LIN	BcFMB_04315	Whole-genome sequencing	Jung et al. (2018)
B. longum (1 isolate)	LIN	*lmrD*	Whole-genome sequencing	Esaiassen et al. (2017)
B. animalis (2 isolates)	TET	*tetT*	Whole-genome sequencing	Esaiassen et al. (2017)
B. breve (1 isolate)				
B. bifidum (Europe strain)	CIP	*glyA*	PCR	Ouoba et al. (2008)
B. longum (Europe strain)	STR	*aadE*		
B. breve strain Yakult	STR	Mutation of *rpsL* gene in G residue at 128 positions	PCR	Kiwaki and Sato (2009)
B. choerinum FMB-1	23S rRNA methyltransferases	BcFMB_02975	Whole-genome sequencing	Jung et al. (2018)
	Aminoglycoside	BcFMB_04310		
B. breve UCC2003	Aminoglycoside	*Bbr_0651*	PCR	Fouhy et al. (2013)
B. breve UCC2003	Aminoglycoside	*Bbr_1586*	PCR	Fouhy et al. (2013)
B. longum subsp. *longum* (1 isolate)	Penicillin	*AIE*	Whole-genome sequencing	Esaiassen et al. (2017)

TABLE 1.1 (Continued)
Antibiotic Resistance Determinants in *Bifidobacterium* Species

Strain	Antibiotic	Representative gene	Genomic analysis	Reference
B. longum subsp. *longum* (1 isolate)	Cefotaxime	*AIE*	Whole-genome sequencing	Esaiassen et al. (2017)
B. breve (2 isolates)	CIP	*mfd, gyrA*	Whole-genome sequencing	Esaiassen et al. (2017)
B. longum subsp. *infantis* (4 isolates)	CIP	*mfd, gyrA*		
B. longum subsp. *longum* (7 isolates)	CIP	*mfd, gyrA*		
B. animalis (2 isolates)	CIP	*mfd, gyrA, gyrB*		

Notes: AMC, aminocoumarin; CIP, ciprofloxacin; FO, Fosfomycin; FQ, Fluoroquinolone; GLY, Glycopeptide; ERY, erythromycin; LIN, lincomycin; MU, Mupirocin; STR, streptomycin; TET, tetracycline; VAN, vancomycin; ND, the result of genes not detected.

et al. (2006), who reported *Lb. rhamnosus*, *Lb. amylovorus*, *Lb. casei*, and *Lb. plantarum* strains in humans and animals.

On the other hand, some lactobacilli discovered the resistance genes. *Cloacal lactobacilli* from broilers were found to have a significant prevalence of CLI resistance (Cauwerts et al., 2006). *Lb. acidophilus* and *Lb. fermentum* resist ERY while *Lb. acidophilus* CRL2152, *Lb. amylovorus* CRL2115, and 4 strains of *Lb. mucosae* (CRL2101, CRL2112, CRL2114, CRL2154) displayed the highest TET resistance (Aristimuño Ficoseco et al., 2018). *Lb. acidophilus*, *Lb. rhamnosus*, and *Lb. casei* from probiotic products exhibited a high prevalence of KAN resistance (Temmerman et al., 2003). Furthermore, resistance to GEN in *Lb. mucosae* CRL2158, resistance to KAN in *Lb. plantarum* CRL2103, resistant to STR in *Lb. acidophilus* CRL2074, *Lb. amylovorus* CRL2065, and *Lb*. mucosae CRL2155 were also discovered (Aristimuño Ficoseco et al., 2018). In agreement, STR resistance in *Lb. acidophilus* from the human origin (Klare et al., 2007) and *Lb. amylovorus* from broilers were reported (Cauwerts et al., 2006).

Although most lactobacilli strains were responsive to antibiotics that suppress protein synthesis, research is controversial about the sensitivity and resistance to the antibiotic of *Lactobacillus* sp. Differences in the examined species, applied methodology, or strain source might cause discrepancies. Because *Lactobacillus* spp. is widely used as a probiotic in food, pharmaceuticals, and beverages, it is vital to investigate the resistance/sensitive and virulence factor genes.

Aristimuño Ficoseco et al. (2018), isolated *Lactobacillus* from feedlot cattle and then determined the virulence gene by using *agg*, *ace*, *espA*, *ebp*, *cylA*, *hyl*, *gelE*, *sprE*, *fsrA*, *fsrB*, *fsrC*, *atpA* as a target gene. However, no feedlot lactobacilli tested positive for virulence factors genes. Contrary, *Lb. salivarius* strain isolated from a sepsis patient was recently discovered to combine human platelets linked to human fibrinogen via discovering a new fibrinogen-binding protein (Collins et al., 2012). Mostly, *Lactobacillus* spp. virulence research has focused on the species *Lb. rhamnosus* and *Lb. paracasei* are the most commonly utilized *Lactobacillus* probiotics (Rossi et al., 2019).

Sirichoat et al. (2020), analyzed the comparative genome sequences against the databases CARD, ResFinder, and ARG-ANNOT. They detected antibiotic-resistant genes: *Lb. jensenii* VA04-2AN resistant to trimethoprim (TMP), *Lb. paracasei* VA02-1AN resistant to CHL and VAN, and *Lb. reuteri* VA24-5 resistant to TMP and VAN. The discovery was the only phenotype and genotype association that was positive. The presence of a conserved phenylalanine residue in the active site of the D-Ala-D-Ala (*Ddl*) ligase, matching amino acid 261 of the *Leuconostoc mesenteroides* enzyme, in the inferred sequence of all vancomycin-resistant strains, while the presence of a tyrosine residue at this location was found in the sensitive strains.

The evidence of antimicrobial resistance and virulence genes in *Lactobacillus* spp. from previous research are summarized in Table 1.2.

1.4.3 *Lactococcus*

Walther et al. (2008) isolated 72 strains of *Lactococcus* from bovine milk, including 41 strains of *Lactococcus lactis* and 31 strains of *Lc. garvieae*, then used a microarray to identify antibiotic-resistance genes. TET, CLI, ERY, STR, and nitrofurantoin resistances were found. Tetracycline-resistant *Lc. garvieae* and *Lc. lactis* carried *tet*(M) and *tet*(S), while clindamycin-resistant *Lc. lactis* contained the *erm*(B) gene and was also resistant to erythromycin. On the other hand, *Lc. lactis* subsp. *lactis* (8 strains) resistance against KAN and TET (Temmerman et al., 2003). In the case of virulence factor genes are not detected in *Lc. lactis* (Bennedsen et al., 2011) (Table 1.2).

TABLE 1.2
Antibiotic Resistance and Virulence Determinants in *Lactobacillus* and *Lactococcus* Strains

Strain	Antibiotic	Resistance genes	Analysis method	Reference
Lb. plantarum KKP 870, KKP 872	CLI	*lnuA*	Disk diffusion tests, PCR-amplifications	Stefańska et al. (2021)
Lb. plantarum KKP 2021p	CLI	*ermB*	Disk diffusion tests, PCR-amplifications	Stefańska et al. (2021)
Lb. casei CRL 2088	ERY	*aph(3")-III, aadA*	PCR-amplifications	Aristimuño Ficoseco et al. (2018)
Lb. plantarum CRL2126	ERY	*cat*	PCR-amplifications	Aristimuño Ficoseco et al. (2018)
Lb. rhamnosus CRL2084	ERY	ND	PCR-amplifications	Aristimuño Ficoseco et al. (2018)
Lb. rhamnosus L-015 and L-455 and *Lb. paracasei* L-005 *Lb. acidophilus* group	STR	ND	Disk diffusion tests, PCR-amplifications	Klare et al. (2007)
Lb. fermentum CRL 2085	STR	*ermB*, *ant*(6), *aadA*	PCR-amplifications	Aristimuño Ficoseco et al. (2018)
Lb. acidophilus CRL2061 and CRL2152	TET	*ermB*	PCR-amplifications	Aristimuño Ficoseco et al. (2018)
Lb. amylovorus CRL2115 and CRL2116	TET	*ermB*	PCR-amplifications	Aristimuño Ficoseco et al. (2018)
Lb. mucosae CRL2064, 2069, 2070, 2083, 2111, 2112, 2114, 2154	TET	*ermB*	PCR-amplifications	Aristimuño Ficoseco et al. (2018)
Lb. fermentum KKP 2020	TET	ND	Disk diffusion tests, PCR-amplifications	Stefańska et al. (2021)
Lb. plantarum (6 strains)	VAN	ND	Disk diffusion tests	Temmerman et al. (2003)

TABLE 1.2 (Continued)
Antibiotic Resistance and Virulence Determinants in *Lactobacillus* and *Lactococcus* Strains

Strain	Antibiotic	Resistance genes	Analysis method	Reference
Lb. acidophilus (13 strains), *Lc. lactis* subsp. *lactis* (8 strains)	KAN	ND	Disk diffusion tests	Temmerman et al. (2003)
Lb. fermentum KKP 811, KKP 830, KKP 843 *Lb. buchneri* KKP 2047p	KAN	*aph(3")-IIIa*	Disk diffusion tests, PCR-amplifications	Stefańska et al. (2021)
Lb. reuteri L-285 and L-285-2	OXY	*tet*(W), *tet*(M)	Disk diffusion tests, PCR-amplifications	Klare et al. (2007)
Lb. mucosae CRL2100 and CRL2101	ERY, TET	*ermB*	PCR-amplifications	Aristimuño Ficoseco et al. (2018)
Lb. mucosae CRL2155	ERY, STR	*tetS*	PCR-amplifications	Aristimuño Ficoseco et al. (2018)
Lb. plantarum CRL2142	ERY, TET	*bla*, *tetS*	PCR-amplifications	Aristimuño Ficoseco et al. (2018)
Lb. acidophilus PCM 2499	STR, TET	*tetK*, *strA/strB*	Disk diffusion tests, PCR-amplifications	Stefańska et al. (2021)
Lb. gallinarum	LIN, CLI	*erm*(B), *erm*(C)	Disk diffusion tests, PCR-amplifications	Cauwerts et al. (2006)
Lb. crispatus L-295 and L-296	ERY, CLI	*erm*(B), *TET*(W)	Disk diffusion tests, PCR-amplifications	Klare Klare et al. (2007)
Lb. amylovorus CRL2044	CLI, TET	*ermB*	PCR-amplifications	Aristimuño Ficoseco et al. (2018)
Lb. rhamnosus (24 strains), *Lb. casei* (29 strains),	KAN, VAN	ND	Disk diffusion tests	Temmerman et al. (2003)
Lb. acidophilus CRL2074	CLI, STR, TET	*ermB*	PCR-amplifications	Aristimuño Ficoseco et al. (2018)
Lb. amylovorus CRL2065	ERY, STR, TET	*ermB*	PCR-amplifications	Aristimuño Ficoseco et al. (2018)
Lb. diolivorans KKP 2036p	KAN, TET, STR	*aph(3")-IIIa*	Disk diffusion tests, PCR-amplifications	Stefańska et al. (2021)
Lb. mucosae CRL2158	ERY, GEN, TET	*tetS*	PCR-amplifications	Aristimuño Ficoseco et al. (2018)
Lb. mucosae CRL2063	CHL, ERY, TET	*ermB*	PCR-amplifications	Aristimuño Ficoseco et al. (2018)
Lb. plantarum CRL2103	CLI, ERY, KAN	*bla*	PCR-amplifications	Aristimuño Ficoseco et al. (2018)
Lb. johnsonii KKP 878	CHL, TET, STR	ND	Disk diffusion tests, PCR-amplifications	Stefańska et al. (2021)
Lb. agilis KKP 1834	GEN, ERY, KAN, STR	*aph(3")-IIIa*, *msrC*	Disk diffusion tests, PCR-amplifications	Stefańska et al. (2021)
Lb. mucosae CRL2113	CLI, CHL, ERY, TET	*ermB*	PCR-amplifications	Aristimuño Ficoseco et al. (2018)
Lb. reuteri (6 strains)	KAN, TET, PEN, VAN	ND	Disk diffusion tests	Temmerman et al. (2003)

(*continued*)

TABLE 1.2 (Continued)
Antibiotic Resistance and Virulence Determinants in *Lactobacillus* and *Lactococcus* Strains

Strain	Antibiotic	Resistance genes	Analysis method	Reference
Lb. jensenii VA04-2AN	TMP	Mutant gene on active site of *Ddl* ligases	Whole-genome sequencing	Sirichoat et al. (2020)
Lb. paracasei VA02-1AN	CHL, VAN			
Lb. reuteri VA24-5	TMP, VAN			
Lc. lactis CHCC6005	TET	*tet*(S)	PCR-amplifications	Bennedsen et al. (2011)
Lc. lactis NZ900	TET, ERY	*Tet*(S) and *erm*(B)	PCR-amplifications	Flórez, et al. (2021)

Notes: AMP, ampicillin; CLI, clindamycin; CHL, chloramphenicol; GEN, gentamycin; ERY, erythromycin; KAN, kanamycin; LIN, lincomycin; OXY, oxytetracycline; STR, streptomycin; TET, tetracycline; TMP, Trimethoprim; VAN, vancomycin; ND, the result of genes not detected. There is no report in virulence genes.

1.4.4 *Enterococcus*

The main point related to antimicrobial resistance is transferring an antimicrobial-resistance gene from bacteria to bacteria, bacteria to an animal, and animal to human via the food chain. Therefore, this problem affects human, animal, medicine, and food production (Dušková and Karpíšková, 2013). *Enterococcus* spp. is a member of the micro-flora organism, which has colonized the digestive system and plays a role in the immune system modification in humans and animals (Krawczyk et al., 2021). They are generally described as Gram-positive, facultatively anaerobic, ovoid-shape with short chains, pair, single-cell, catalase-negative bacteria (Mohanty and Bhuniya, 2021). They were isolated from many sources, comprising soil, surface water, plants, and animals. In the last decade, this bacterial group has been widely used as probiotics and starter cultures in the food industries. In contrast, they have recently been described as the nosocomial pathogen that causes intensive care infections. Significantly, both *Enterococcus faecium* and *Enterococcus faecalis* are important surveillance pathogens because they are able to resist multiple drugs (Mohanty and Bhuniya, 2021). Vancomycin resistance was mediated by nine distinct gene clusters, including *vanA, B, C, D, E, G, L, M, and vanN*, found (Hollenbeck, and Rice, 2012). De Leener et al. (2004) identified macrolide resistance genes as *ermB* and TET resistance genes as *tetK, tetL, tetM, tetO*, and *tetS* from *Enterococcus* strains isolated from carcasses of humans, pigs, and pork. The antibiotic resistance of *Enterococcus* spp. represents the β-lactam resistance gene TEM, aminoglycoside resistance genes *aac(6')-aph(2")*, *aph(3')-III*, *ant(6)-I* and *ant(2")-I*, the TET resistance gene *tetM*, erythromycin resistance gene *ermB*, the vancomycin resistance gene *vanA*, and the enterococcal multidrug resistance efflux pump gene *emeA*, which are determined by a PCR (Jia et al., 2014). Raeisi et al. (2017) used disk diffusion and polymerase chain reaction (PCR) to identify the expression of the vancomycin resistance gene as a *vanA, gene in Enterococcus faecalis.* The DNA microarray is the technique that detects the medium-length oligonucleotides by using the target probe detection (Missoum, 2018). Diarra et al. (2010) examined the antimicrobial resistance genes in enterococci. Moreover, this experiment determined the genes resisting aminoglycoside (*aac, aacA-aphD, aadB, aphA, sat4*), macrolide (*ermA, ermB, ermAM, msrC*), TET (*tetL, tetM, tetO*), streptogramin (*satG_vatE8*), bacitracin (*bcrR*), and LIN (*linB*) by utilization of DNA microarray.

Among enterococci species widely found in nosocomial infections as pathogens, included *Enterococcus faecalis* and *Enterococcus faecium.* In addition, *Enterococcus* spp. contain several virulence genes. The virulence genes determination was investigated using the PCR technique, which uses the specific primers for amplification. They found virulence genes, including the *E. faecalis* endocarditis antigen (*efaA*) gene and enterococcal surface protein (*esp*) gene. In contrast, they did not find the aggregation substance (*AS*) gene, collagen-binding protein (*ace*) gene, cytolysin (*cylA*) gene, gelatinase (*gelE*) gene, and hyaluronidase (*hyl*) gene in *E. faecium* of this study. The *asa* gene is encoded into a surface protein, which involves the conjugative transferability of the plasmid. *ace*, *efaA*, *esp* are proteins, which increase the adhesion and biofilm formation of the pathogen to other surfaces. The gelatinase encoded by *gelE* has rolled in the gelatin hydrolysis. The expression of the *esp* gene extensively induces this bacterial strain's conjugation (Mannu et al., 2003). Additionally, Billström et al. (2008) reported the possibility of the correlations between virulence and antimicrobial resistance. They demonstrated *esp* gene expression correlates with ampicillin, CIP, and imipenem resistance ($P< 0.01$). They determined seven virulence genes (*asa1*, *cylA*, *ace*, $efaA_{fs}$, esp_{fm}, *gelE*, and hyl_{fm}) in *E. faecium* by PCR method. Among 263 isolates of *E. faecium*, 56% represent *espfm*, which present enterococcal surface protein. 11 isolates present $hyl_{fm,}$ which encoded hyaluronidase. Moreover, different antimicrobial resistance and virulence genes in *Enterococcus* spp. as shown in Table 1.3.

TABLE 1.3
Antibiotic Resistance and Virulence Genes Determinants in *Enterococcus* Species

Strain	Antibiotic	Resistance genes	Virulence genes	Analysis method	Reference
E. faecalis group	β-Lactam	*TEM*	Not report	PCR	Jia et al. (2014)
	Aminoglycoside	*aac(6')/aph(2')* *aph(3')-III* *ant(6)-I* *ant(2")-I*			
		aac6	Collagen-binding cell wall protein (*ace*) cytolysin gene (*cylA*, *cylB*) gelatinase gene (*gelE*)	Microarray	Diarra et al. (2010)
		aphA3	endocarditis-specific antigen ($efaA_{Efs}$) Aggregative pheromone-inducing adherence to extra-matrix protein (*agg*) AgrB protein ($agrB_{Efs}$) pheromone cAD1 precursor lipoprotein (*cad1*) sex pheromone cAM373 precursor (cAM373) Pheromone cCF10 precursor lipoprotein (cCF10) pheromone cOB1 precursor/lipoprotein (*cob*)		
	TET	*tetL* *tetM*	ND	Microarray	Diarra et al. (2010)
		tetM	Not report	PCR	Jia et al. (2014)

(*continued*)

TABLE 1.3 (Continued)
Antibiotic Resistance and Virulence Genes Determinants in *Enterococcus* Species

Strain	Antibiotic	Resistance genes	Virulence genes	Analysis method	Reference
		tetK *tetL* *tetM* *tetO* *tetS*	Not report	PCR	De Leener et al. (2004)
	Macrolide or erythromycin	*ermAM* *ermB*	Not report	Microarray	Diarra et al. (2010)
		ermB	Not report	PCR	Jia et al. (2014)
		ermB	Not report	PCR	De Leener et al. (2004)
	Multidrug resistance efflux pump	*emeA*	Not report	PCR	Jia et al. (2014)
	KAN	*aphA3*	ND	Microarray	Diarra et al. (2010)
	Mercury resistant	*mer*	Not report	PCR DNA-DNA hybridization	Soge et al. (2008)
E. faecium group	β-lactam-resistant	*TEM*	Not report	PCR	Jia et al. (2014)
	Aminoglycoside	*aac(6')/aph(2')* *aph(3')-III* *ant(6)-I* *ant(2")-I*	Not report	PCR	Jia et al. (2014)
		aac6 *aacA-aphD1* *aad9* *aphA3* *aadB*	*E. faecalis* specific cell wall adhesin (*efmA*$_{Efm}$)	Microarray	Diarra et al. (2010)
	TET	*tetL* *tetM*	ND	Microarray	Diarra et al. (2010)
		tetL	Not report	PCR	De Leener et al. (2004)
		tetM	Not report	PCR	Jia et al. (2014)
		tetM *tetS*	Not report	PCR	De Leener et al. (2004)
	Macrolide or ERY	*ermA* *ermAM* *ermB* *msrC*	ND	Microarray	Diarra et al. (2010)
		ermB	Not report	PCR	Jia et al. (2014)
		ermB	Not report	PCR	De Leener et al. (2004)
	Glycopeptide or VAN	*vanA*	Not report	PCR	Jia et al. (2014)
	Multidrug resistance efflux pump	*emeA*	Not report	PCR	Jia et al. (2014)
	KAN	*aphA3*	ND	Microarray	Diarra et al. (2010)
	Mercury resistant	*mer*	Not report	PCR DNA-DNA hybridization	Soge et al. (2008)

TABLE 1.3 (Continued)
Antibiotic Resistance and Virulence Genes Determinants in *Enterococcus* Species

Strain	Antibiotic	Resistance genes	Virulence genes	Analysis method	Reference
	Streptogramin	*satG_vatE*	ND	Microarray	Diarra et al. (2010)
	LIN	*linB*	ND	Microarray	Diarra et al. (2010)
E. hirae group	Aminoglycoside-resistant *Enterococcus*	*aac6* *aacA-aphD1* *aad9* *aphA3*	Pheromone cAD1 precursor lipoprotein (*cad1*)	Microarray	Diarra et al. (2010)
	TET	*tetL* *tetM* *tetO*	ND	Microarray	Diarra et al. (2010)
		tetL *tetM*	Not report	PCR	De Leener et al. (2004)
	ERY	*ermAM* *ermB*	ND	Microarray	Diarra et al. (2010)
		ermB	Not report	PCR	De Leener et al. (2004)
	KAN	*aphA3*	ND	Microarray	Diarra et al. (2010)
	STR	*satG_vatE*	ND	Microarray	Diarra et al. (2010)
	LIN	*linB*	ND	Microarray	Diarra et al. (2010)
	Mercury	*mer*	Not report	PCR DNA-DNA hybridization	Soge et al. (2008)
E. gallinarum group	Aminoglycoside	*aacA-aphD1*	ND	Microarray	Diarra et al. (2010)
	TET	*tetL* *tetM*	ND ND	Microarray	Diarra et al. (2010)
		tetM	Not report	PCR	De Leener et al. (2004)
	Macrolide or EYR	*ermAM* *ermB*	ND	Microarray	Diarra et al. (2010)
	Glycopeptide or VAN	*vanC*	ND	Microarray	Diarra et al. (2010)
	Mercury	*mer*	Not report	PCR DNA-DNA hybridization	Soge et al. (2008)
E. avium group	TET	*tetL* *tetK* *tetM*	Not report	PCR	De Leener et al. (2004)
	Macrolide or ERY	*ermB*	Not report	PCR	De Leener et al. (2004)

Notes: ERY, erythromycin; KAN, kanamycin; LIN, lincomycin; STR, streptomycin; TET, tetracycline; VAN, vancomycin; ND, the result of genes not detected.

1.4.5 *Pediococcus*

The resistant genes; *tetK*, *tetL*, *tetM*, *tetO*, *tetS* (tetracycline resistance), *catA* (chloramphenicol resistance), *aadE* (aminoglycoside STR), *erm* (erythromycin), and *genR* (gentamicin *resistance*) are frequently used to determine for antimicrobial resistance gene in *Pediococcus* spp. (Danielsen et al., 2007; Klare et al., 2007; Zommiti et al., 2018).

Previous research has correlated *Pediococcus* sp. antibiotic resistance/susceptibility to various results. For instance, *P. acidilactici* (29 strains) and *P. pentosaceus* (20 strains) were isolated from human origin, sensitive to STR, VAN, EYR, CLI, OXY, and fusidic acid (Klare et al., 2007). When *P. acidilactici* CRL2043 and CRL2046 from feedlot cattle were examined, resistance to ERY and TET was found, but sensitivity to CLI, CHL, GEN, KAN, and STR was also discovered. The research of Danielsen et al. (2007) reported that *P. acidilactici*, *P. damnosus*, *P. dextrinicus*, *P. parvulus*, *P. inopinatus*, and *P. pentosaceus* were susceptibile to AMP, CHL, GEN, and STR. Nonetheless, a gene for ERY resistance, *erm*(B), was detected by a PCR in ERY-resistant *P. acidilactici* and *P. pentosaceus*. Furthermore, *P. pentosaceus* was naturally resistant to both TET and VAN (Danielsen et al., 2007). Zommiti et al. (2018) reported that *P. pentosaceus* MZF16 was resistant to GEN, chloramphenicol (CHL), and VAN. *P. pentosaceus* LPM78 resists kanamycin, erythromycin, and clindamycin (Muñoz-Atienza et al., 2013). *P. pentosaceus* LPM78 contains the erythromycin resistance gene *mef*(A/E). In addition, the *lnu(A)* gene, which encodes LIN O-nucleotidyltransferase, and inhibits lincomycin and clindamycin, is also found in *P. pentosaceus* LPM78 and LPM83. However, *lnu*(B) was not found in any examined strains (Muñoz-Atienza et al., 2013).

In addition, the antibiotic susceptibility effect was consistent with the *P. pentosaceus* strains, including strains KCC-23, ST65ACC, LJR1, KID7, CRAG3, and MZF16. Nevertheless, some differences in the resistance and virulence genes in the *Pediococcus* strains are described in Table 1.4.

P. pentosaceus MZF16 is resistant to VAN strain (Zommiti et al., 2018). The result is supported by the report that VAN often exists in *Pediococcus* sp. (Toomey et al., 2010). Nonetheless, intrinsic VAN resistance results from a modified peptidoglycan precursor ending in D-Ala-D-lactate, leading to the absence of horizontal transmission in non-pathogenic microorganisms and posing no risk (Ammor et al., 2007).

For detection of virulence genes, the PCR amplification technique and genome analysis are mostly used for encoding the protein that causes aggregation, including (*agg*), gelatinase (*gelE*), enterococcal surface protein (*esp*), endocarditis and biofilm-associated pilus (*ebp*), accessory colonization factor (*ace*), cytolisin (*cyl*A), hyaluronidase (*hyl*), gelatinase (*gelE*), serineprotease (*sprE*), but in *Pediococcus* spp. the virulence factor is absent. For examples of research claims that virulence factors are absent in *Pediococcus* spp. see Borges et al., 2013and Muñoz-Atienza et al., 2013). Cavicchioli et al. (2019) reported that *P. pentosaceus* ST57ACC involved in histamine, tyramine, and putrescine gene expression is absent. However, Lüdin et al. (2018) explained antibiotic-resistant genes using genome assembly. The result investigated that *P. acidilactici* FAM 13875 (GenBank accession GCA_009789085.1) consisted of antibiotic-resistance genes *tetM* and *ermA*. Furthermore, the lack of virulence factor requires microorganisms to be classified as probiotics (Table 1.4).

1.5 MECHANISMS OF ANTIBIOTIC RESISTANCE

The antibacterial resistance mechanisms are divided into four groups: restricted absorption of antimicrobial agents, target modification, drug inactivation by enzyme digestion, and drug efflux (Reygaert, 2018). Gram-negative bacteria adopt all four mechanisms, but Gram-positive bacteria commonly adopt limiting drug absorption only (Chancey et al., 2012; Mahon et al., 2014). The common antimicrobial resistance mechanisms are illustrated in Figure 1.3.

The microbial susceptibility and resistance measurements use minimum inhibitory concentration (MIC), the minimal antimicrobial agent concentration that inhibits bacterial growth. There are two different resistances in microorganisms: native resistance and acquired resistance. Native

TABLE 1.4
Antibiotic Resistance and Virulence Genes Determinant in *Pediococcus* Species

Strain	Antibiotic or inhibitor	Resistance gene	Analysis method	Reference
P. acidilactici (8 strains)	VAN	ND	Disk diffusion tests	Temmerman et al. (2003)
P. acidilactici CRL2043, 2046	ERY, TET	ND	PCR-amplifications	Aristimuño Ficoseco et al. (2018)
P. acidilactici and *P. pentosaceus*	ERY, TET, VAN	*erm*(B)	Etest	Danielsen et al. (2007)
P. pentosaceus KCC-23	GEN, CN, COT, AK, Colistin methanesulphonate, SF	Not report	Disk diffusion tests	Ilavenil et al. (2016)
P. pentosaceus ST65ACC	Van, Trimethoprim/ sulfamethoxazole, Oxacillin	*ccf*	Disk diffusion tests, PCR-amplifications	Cavicchioli et al. (2019)
P. pentosaceus LJR1	Group of cell wall synthesis inhibitor, Nucleic acid synthesis inhibitor, Protein synthesis inhibitor	Not report	Disk diffusion tests	Ladha and Jeevaratnam (2018)
P. pentosaceus KID7	STR, KAN	ND	Disk diffusion tests, PCR-amplifications	Damodharan et al. (2015)
P. pentosaceus CRAG3	CIP, KAN, VAN	Not report	Disk diffusion tests	Shukla and Goyal (2014)
P. pentosaceus MZF16	GEN, CHL, VAN	ND	Disk diffusion tests, PCR-amplifications	Zommiti et al. (2018)
P. pentosaceus LPM78	ERY, ERY, CLI, LIN	mef(A/E), *lnu*(A)	Disk diffusion tests, PCR-amplifications	Muñoz-Atienza et al. (2013)
P. pentosaceus LPM83	LIN, CLI	*lnu*(A)	Disk diffusion tests, PCR-amplifications	Muñoz-Atienza et al. (2013)
P. acidilactici FAM 13875	TET, ERY	*tetM*, *ermA*	Whole-genome sequencing	Lüdin et al. (2018)

Notes: AK, amikacin, CHL, chloramphenicol; CLI, clindamycin; CN, cefalexin; COT, co-trimoxazole; GEN, gentamicin; ERY, erythromycin; KAN, kanamycin; LIN, lincomycin; SF, sulphafurazole, STR, streptomycin; TET, tetracycline; VAN, vancomycin; ND, the result of resistance not detected. There are no report in virulence genes.

resistance is divided into two ways, intrinsic and induced resistance. Intrinsic resistance is the resistance that usually occurs in a bacterial species. The induced resistance is expressed after receiving an antimicrobial. The intrinsic resistance mechanisms are generally found to reduce outer-membrane permeability and induction of efflux drugs. The induced resistance typically found in bacteria is multidrug efflux. The gene mutation causes acquired resistance via gene transformation, transposition, and conjugation. Moreover, external factors such as antimicrobial use, UV radiation, starvation, chemical, etc., lead bacteria to mutate the gene, increasing antimicrobial resistance (Reygaert, 2018). The mechanisms of antibiotic resistance in LAB are presented in Table 1.5.

Tetracyclines are described as antimicrobials, which inhibit bacterial growth and protein synthesis. These tetracyclines act on bacterial cells but not on mammalian cells. Tetracyclines pass through bacterial outer-membrane and inner-membrane by passive diffusion and energy-dependent active transport, respectively. Further, they bind to the ribosome and block the entrance of aminoacyl-tRNA to the ribosomal receptor. As for results, tetracyclines interrupt the membrane permeability caused by cell damage, lysis, and death. Tetracycline resistance refers to the tetracycline efflux-mediated mechanism or protection of the ribosomes from the action of tetracycline. These proteins

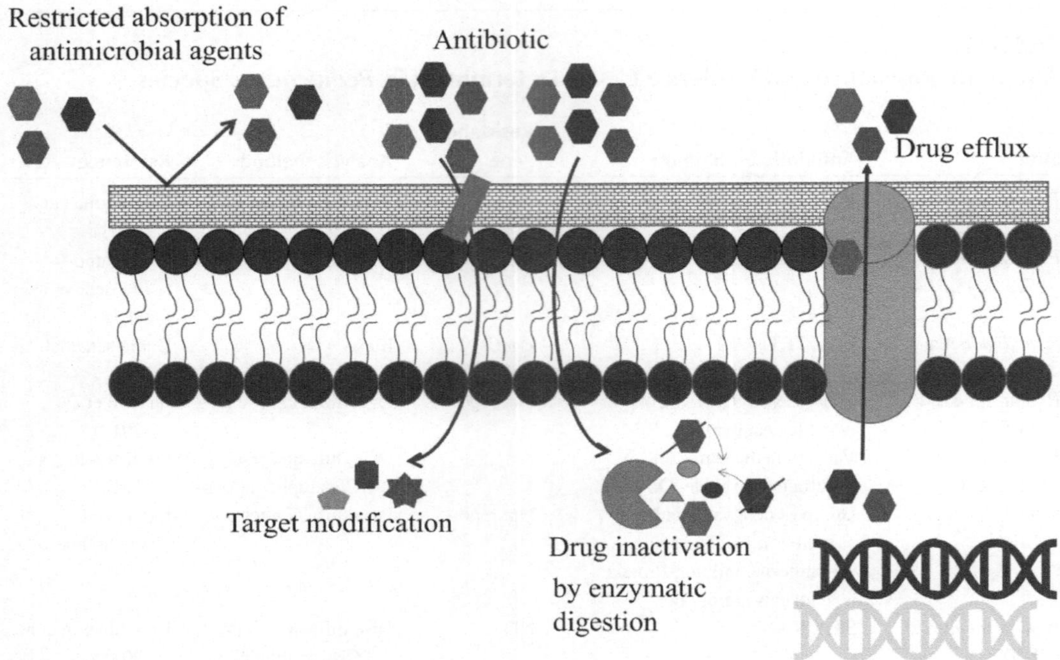

FIGURE 1.3 Common mechanisms of resistance to antimicrobial agents in LAB.
Source: Modified from Álvarez-Cisneros and Ponce-Alquicira (2018).

TABLE 1.5
Mechanisms of Antibiotic Resistance in LAB

Mechanism	Antibiotic	Genus	Reference
Resistance absorption of antimicrobial agent	Glycopeptides	*Enterococcus*	Rice (2012)
Target modification	Tetracyclines	*Bifidobacterium*	Kim et al. (2018)
		Enterococcus	De Leener et al. (2004)
	Macrolide Lincosamide Streptogramin	*Pediococcus*	Danielsen et al. (2007)
Drug inactivation by enzymatic digestion	β-Lactam	*Enterococcus*	Rice (2012)
	Macrolide Lincosamide	*Lactobacillus*	Cauwerts et al. (2006)
	Aminoglycoside	*Enterococcus*	Padmasini et al. (2014)
Drug efflux	Tetracyclines	*Enterococcus*	De Leener et al. (2004)
	Tiamulin	*Enterococcus*	Wendlandt et al. (2013)
		Lactobacillus	Dec et al. (2017)
	Macrolide Lincosamide	*Lactobacillus*	Cauwerts et al. (2006)

are encoded by *tet* gene (*tetA*, *tetB*, *tetC*, *tetD*, *tetE*, *tetG*, *tetH*, *tetK*, and *tetL*) to export and decrease the concentration of tetracyclines in the cell by efflux mechanism. Additionally, expressions of ribosomal protected protein increase the tetracycline resistance of bacteria. *Bifidobacterium longum* BORI has tetracycline W (*tetW*), encoded the factor GTPase of the TRAFAC. This factor increases the modification of bacterial ribosomes, decreasing the tetracycline activity by inhibiting protein synthesis (Kim et al., 2018). De Leener et al. (2004) demonstrated tetracycline resistance genes *tetK*, *tetL*, *tetM*, *tetO*, and *tetS* in *Enterococcus* strains, representing involved tetracycline resistance mechanisms.

β-Lactam antimicrobial interferes with peptidoglycan cell wall production by binding penicillin-binding proteins (PBPs) as the cell wall production enzyme. This reaction leads to inhibiting PBPs, causing a reduction of the bacterial cell wall and bacterial growth and cell death. β-lactam-resistant is described as the reduction of PBP binding accession, decreasing of PBP binding affinity, and degradation of β-lactams mechanisms via expression of β-lactamase, which is a β-lactams degradable enzyme (Lima et al., 2020).

Glycopeptides are adequate size peptides that can pass through Gram-negative bacterial cell walls via porins. Glycopeptides bind the terminal D-alanine position of pentapeptide of peptidoglycan precursors. As for results, they inhibit the transpeptidation process necessary for peptidoglycan cross-linking. Glycopeptide resistance is caused by the expression of operons that replace the D-alanine with D-lactate or D-serine at the pentapeptide end-terminal. The operons involved in vancomycin resistance are included with *vanA*, *vanB*, *vanC*, *vanD*, *vanE*, *vanG*, *vanL*, *vanM*, and *vanN* gene (Ahmed and Baptiste, 2018). Therefore, the binding affinity of vancomycin is decreased, which demonstrates the vancomycin-resistant (Rice, 2012).

Tiamulin is the semisynthetic derivative of the pleuromutilin compound. These active compounds inhibit protein production via 50S ribosomal subunit and peptidyl transferase attractions (Borak et al., 2020). The ABC transporters roll to increase the drug efflux mechanism. *lsaE* represent in *Enterococcus* and *Lactobacillus* strains and encoded into ABC *lsa*(A) protein, which is involved with pleuromutilin (tiamulin) resistance (Dec et al., 2017; Wendlandt et al., 2013).

Macrolide, LIN, and streptogramin are categorized in the same group based on modes of action compound. Macrolides are considered lactones and attached amino, which have antimicrobial activity. They disrupt the protein synthesis of the peptidyl-tRNA molecule from the ribosomes during elongation. As for the macrolide, STR, and LIN resistance by target modification, adenine-N6 methyltransferase, encoding by *erm* add a methyl group to the 23S rRNA as a posttranscriptional modification. Therefore, these bacteria are resistant to macrolides, lincosamides, and streptogramin B antibiotics (MLS_B) (Cattoir and Leclercq, 2017). The expression of *ermB* of *P. acidilactici* NCIMB 6990 increased the MIC of erythromycin agent higher than 256 μg/ml (Danielsen et al., 2007). Furthermore, the macrolide, lincosamide, and streptogramin resistance in bacteria involves gene expressions of efflux genes and the inactivation of antimicrobials. Representation of *mefA* and *lmrA* produce transport proteins, and encoding *carA*, *msrA*, *oleB*, and *srmB* produces ATP-binding transporters. These gene expressions induce the antibiotic efflux of bacteria. The inactivation of antimicrobials relates to enzyme encoding genes: *ereA* and *ereB* (esterases); *lnuA* and *lnuB* (transferases); *mphA* and *mphB* (phosphorylases). *Lactobacillus* species isolated from broilers present *ermB*, *ermC*, *InuA*, and *mefA* genes, which involve tylosin, erythromycin, and lincomycin resistance phenotypes (Cauwerts et al., 2006).

Streptogramins are the natural cyclic peptides consisting of group A and group B streptogramins. Both streptogramins act on 50S ribosomal subunits at the peptidyl transferase domain. They block the translation of messenger RNA (mRNA) at the peptide chain elongation process. There are three different mechanisms of streptogramin resistance: modification, inactivation, and efflux increasing of the antimicrobials. The methylation of ribosomal RNA induces group B streptogramin (Cattoir and Leclercq, 2017).

Aminoglycoside resistant, gene expression of *aac(6')/aph(2')*, *aph(2'')-Ib*, *aph(2'')-Ic, and aph(2'')-Id*, which produces bifunctional aminoglycoside modifying enzyme AAC(6')-APH(2'') lead high-level gentamicin resistance. This enzyme decreases the synergistic gentamycin activity when it integrates with the cell wall active agent (ampicillin or vancomycin) (Padmasini et al., 2014).

Antibiotics and probiotics are widely utilized to improve human and animal health. However, LAB are considered the probiotics that carry antimicrobial resistance genes, which are able to transfer from bacteria to humans via the food chain. The translocation of DNA elements may initiate with the unawareness use of antibiotic treatment (Álvarez-Cisneros and Ponce-Alquicira, 2018). Three different mechanisms of horizontal gene transfer are comprised of conjugation, transformation, and transduction. (1) Conjugation is the direct DNA transmission between the donor and recipient bacteria (Graf et al., 2019). As for the conjugation of LAB, Jacobsen et al. (2007) reported the trans-conjugations of *tetM* and *ermB* resistance elements from *Lb. plantarum* isolated from fermented food to *E. faecalis.* (2) Transformation is the environmental DNA receiving of bacteria. After cell death, the genomic element may be released into the environment. The surrounding bacteria take up DNA elements in the cell. The survival and incorporation of DNA initiate the expression of acquired DNA in the bacterial cell. However, the naturally transforming antimicrobial resistance genes from food to LAB have not been reported. they might be rare situations or difficult detections. (3) Transduction is the mechanism that uses bacteriophage as the carrier. In this case, the target bacteria are attached and receive the DNA element from the bacteriophage. The stabilized genetic fragment either auto-replicates or integrates with the host DNA.

Thus, new bacteriophages can be produced within the bacteria, and the phage genetic element can be transferred to others also. On the other hand, the antimicrobial resistance gene transfer of LAB via the transduction mechanism is still infrequently reported as the transformation (Verraes et al., 2013).

1.6 EVALUATION OF ANTIBIOTIC RESISTANCE AND THE REGULATION OF THE USE OF LAB AS PROBIOTICS

Antimicrobial susceptibility is generally detected by two methods: antimicrobial resistance phenotypic and genotypic characteristics. The minimum inhibitory concentrations (MICs) are typically used to evaluate the antimicrobial susceptibility of LAB as phenotypic characteristics. Broth microdilution is used to determine the MICs of LAB except for the *Enterococcus* species (Álvarez-Cisneros and Ponce-Alquicira, 2018). The MICs determination of *Enterococcus* strains used the Sensititre automated system, described by the Clinical and Laboratory Standards Institute (CLSI) (Diarra et al., 2010). Examples of MICs examined techniques include broth microdilution (MDIL), Etest, and the Kirby-Bauer test (Biemer, 1973; Kushiro et al., 2009). The advantage of the MDIL method is the MICs examination of a great many isolates and antibiotics. Thus this method is widely used in research (Álvarez-Cisneros and Ponce-Alquicira, 2018). The cut-off value is used to evaluate MICs or antimicrobial susceptibility, followed by Laulund et al. (2017). Furthermore, the observation of antimicrobial-resistant genes of LAB as the genotypic characteristics is typically used poly chain reaction (PCR) techniques, microarrays, and whole-genome sequencing (Diarra et al., 2010; Jia et al., 2014; Rozman et al., 2020).

Several genomic analysis databases are utilized to predict antimicrobial resistance genes, gene mutations, and translocation of genetic elements. Moreover, the metagenomic analysis increases the understanding of antimicrobial resistance mechanisms, which involve bacteria and other factors such as food (Álvarez-Cisneros and Ponce-Alquicira, 2018). In addition, Zheng et al. (2017) suggested that computer models are also used to represent adaptive prediction and metamorphic approaches of commercially produced probiotic bacteria. The model could be used to evaluate the risk of probiotic dietary supplements from the perspective of antimicrobial resistance. Designed to assist manufacturers and medical practitioners, as well as to provide alternative information to consumers about the advantages and risks of using probiotic dietary supplements. The overarching objective is to allow probiotic bacteria with numerous promising health advantages to be utilized responsibly and with little danger to long-term human health.

Antibiotic resistance in microorganisms used as animal feed additives must be regulated to prevent strains carrying highly infectious antibiotic resistance genes from entering the food chain and spreading resistance to human or animal infections. In practice, most bacteria are resistant to certain antibiotics (Rabello et al., 2020). The transmission of resistance to vulnerable organisms representing other species or genera is exceedingly uncommon if the resistance is inherent or dependent on the strain's physiological or structural properties (cell wall characteristics, absence of antibiotic target function, etc.) (Wright, 2005). The phenotypic of vancomycin-resistant enterococci is based on transmissible genetic factors (Ahmed and Baptiste, 2018), whereas certain lactobacilli exhibit intrinsic vancomycin resistance owing to peptidoglycan structural variations (Nayyar et al., 2022). Of course, both situations have clear safety concerns.

As a result, when using microorganisms in food, there should be criteria that indicate the level of resistance to an antibiotic in probiotic limit for consumer safety and regulate their spread in the environment. The distribution of minimum inhibitory concentrations (MICs) of antimicrobials in bacterial populations of the same species or genus is used to determine cut-off values (Álvarez-Cisneros and Ponce-Alquicira, 2018).

The Food and Agriculture Organization (FAO) and the World Health Organization (WHO) published a report in 2002 related to the evaluation of probiotic properties. Since probiotic benefits are strain-specific, it is necessary to identify the strains using phenotypic and genotypic approaches to be categorized as probiotics. Bacteria to be used as probiotics should undergo functional characterization and safety assessments both in vitro and in animal tests before subsequent examination for efficacy in human research that was double-blind, randomized, and placebo-controlled (WHO, 2002). A strain must first pass a second trial standard before being declared a probiotic and added to a food product (Binda et al., 2020).

Based on the minimal evidence provided, the MIC values for LAB appear to be appropriate. For example, the European Food Safety Authority (EFSA) used publicly available information on the levels of antibiotic resistance as antimicrobials important to humans and veterinarians, including antimicrobial agents such as *Bifidobacterium*, *Enterococcus*, *Lactobacillus*, *Lactococcus*, *Pediococcus*, *Streptococcus thermophilus*, and *Bacillus* species. Susceptibility to these substances is a basic requirement for bacterial products intended for use as food additives and animal feeds while developing its safety evaluation method (Zhou et al., 2012).

In addition, the European Union has also developed a comprehensive framework for defining microorganism resistance in *Lactobacillus* spp. for use as a food and feed additive (Salvetti and O'Toole, 2017). However, data on MIC breakpoints of antibiotic-resistant are limited. Therefore, for the bacteria most often used as probiotics, such as *Bifidobacterium*, *Enterococcus*, *Lactobacillus*, *Lactococcus*, and *Pediococcus*, resistance limit criteria are shown in Table 1.6.

Probiotics should not contain more resistance than is required to prevent possible resistance transfer into pathogenic bacteria (Patel et al., 2012). The Scientific Committee on Animal Nutrition (SCAN) also notes that the MIC values presented constitute a somewhat restricted practice recommendation in identifying the equivalent values at or beyond which the occurrence of transmissible antibiotic resistance should be investigated. In conclusion, If the MIC value for a particular antibiotic resistance is exceeded, conjugation studies should be used to assess the ability to transmit impedance. If no gene transfer is identified, such strains should be tested for the presence of known antibiotic resistance genes using the PCR technique. In case of the absence of identified resistance genes, the intrinsic or mutational nature of antibiotic resistance should be determined. In practice, this last need constitutes a hard challenge. It is possible to identify and isolate the genes responsible for the phenotype (Wright, 2005).

At present, attempts are being undertaken to employ alternative antimicrobials to reduce the problem of antibiotic resistance and distribution in the environment. According to studies by Czaplewski et al. (2016) and Seal et al. (2018), vaccination, biosecurity, good animal husbandry programs, prebiotics, antimicrobial peptides, bacteriophages, gene products, antibodies, and natural

TABLE 1.6
Antibiotic Resistant Breakpoints for *Bifidobacterium, Enterococcus, Lactobacillus, Lactococcus,* and *Pediococcus* as Probiotics in Food Products

	Microbiological breakpoint, MIC (μg/ml)				
Antibiotic	***Bifidobacterium***	***Enterococcus***	***Lactobacillus***	***Lactococcus***	***Pediococcus***
Ampicillin	2[a]	2[a] 8[d]	1[a,c], 2[b]	2[a]	3[d], 4[a]
Chloramphenicol	4[a]	16[a,d]	4[a,c], 16[b]	8[a]	4[a], 16[d]
Chlortetracycline	NR	NR	1[c]	NR	NR
Cipro/enrofloxacin	NR	2[d]	NR	NR	4[d]
Clindamycin	1[a]	4[a]	1[a]	1[a]	1[a]
Erythromycin	1[a]	4[a,d]	1[a]	1[a]	1[a], 4[d]
Gentamicin	4[a]	32[a], 500[d]	16[a,c], 1[b]	32[a]	16[a], 4[d]
Kanamycin/neomycin	8[a]	1024[a,d]	16[a,c]	64[a]	32[d], 64[a]
Lincomycin	NR	NR	4[c]	NR	NR
Linezolid	NR	4[d]	4[a]	NR	4[d]
Neomycin	NR	NR	32[c]	NR	NR
Penicillin G	NR	NR	1[c]	NR	NR
Quinupristin/dalfopristin	4[a]	4[d]	16[a]	4[a]	4[a,d]
Rifampin	NR	4[d]	NR	NR	4[d]
Roxithromycin	NR	NR	1[c]	NR	NR
Streptomycin	8[a]	128[a], 1024[d]	64[a,c]	32[a]	32[d], 64[a]
Tetracycline	8[a]	4[a], 16[d]	4[a,c], 16[b]	4[a]	8[a], 16[d]
Trimethoprim	NR	8[d]	NR	NR	32[d]
Vancomycin	2[a]	4[a,d]	4[c]	4[a]	4[d]

Notes: The information of Breakpoints from [a]European Food Safety Authority (EFSA, 2012), [b]European Commission (European Commission, 2003), [c]Zhou et al. (2012), and [d]Wright (2005); NR, not required.

substances such as polyphenols and especially probiotics are expected to replace antimicrobial growth promoters in the livestock sector.

In addition, lactic acid bacteria are beneficial probiotics as another choice to reduce the use of antibiotics in livestock. Since they could change the environment in which they are administered by creating a variety of metabolites, including a diverse set of inhibitory chemicals, as well as competitive exclusion (Gaggìa et al., 2010). Production of inhibitory chemicals, avoidance of pathogen adherence, competition for nutrients, modification of the host immune system, enhancement of nutrient digestibility, feed conversion, and decrease of toxin bioavailability are some of the mechanisms of pathogen suppression by LAB-probiotics (Vieco-Saiz et al., 2019). Moreover, numerous phylogenetic studies related to experimental data have established host-adaptation paradigms (Frese et al., 2011).

The food processing industry's attempt to control antimicrobial resistance of LAB by using an appropriate starter, and an acceptable substrate for fermentation, such as pasteurized milk, are used in combination with antibiotics used with caution. One strategy is to prevent the spread of microorganisms that cause antibiotic resistance (Amiri et al., 2021; Hashempour-Baltork et al., 2019). Similarly, Patel et al. (2012) suggested that the preventing and regulating strategies will require the application of epidemiological and behavioural approaches, such as investigating the mechanism of antimicrobial resistance. The use of LAB and its metabolites as a natural curative agent should be promoted to limit the spread of resistance genes gene transfers in both clinical and food-grade bacteria (Patel et al., 2012). In addition, Fraqueza (2015) described that the primary species in the microbiota of dry-fermented sausages were *Lb. sakei* and *Lb. plantarum* and special attention should be paid to antibiotic resistance profiles before they are used as starters in the production of dry-fermented sausages because some antibiotic resistances are expressed in LAB without the identification of recognized genes. The significance of creating approaches that adapt to different nucleotide sequences of the resistance genes is emphasized and other gene tests indicate the same antibiotic resistance gene. Additionally, differences in resistance phenotypes could be due to inactive or absent genes; therefore, antibiotic-resistant LAB in dry-fermented sausages should be minimized or avoided. Several factors contribute to this issue. As a result, a combination of environmental and antibiotic consumption in animal feed and humans should be managed (Fraqueza, 2015).

1.7 CONCLUSION AND FUTURE PROSPECTS

The large-scale antibiotic application in human and veterinary medicine, agriculture, industry, and aquaculture is pointed out as one of the main causes of the increasing rate of resistant bacteria. At present, attempts are being undertaken to employ alternative antimicrobials to reduce the problem of antibiotic resistance and distribution in the environment. Vaccination, biosecurity, good animal husbandry programs, prebiotics, antimicrobial peptides, bacteriophages, gene products, antibodies, and natural substances are expected to replace antimicrobial growth promoters in the livestock sector. LAB are beneficial probiotics as one another choice to reduce the use of antibiotics in livestock because they have the potential to alter the environment in which they are delivered by producing a wide range of metabolites. Some of the mechanisms of pathogen suppression by LAB-probiotics include avoidance of pathogen adherence, competition for nutrients, change of the host immune system, augmentation of nutrient digestibility, feed conversion, and decrease of toxin bioavailability.

Notwithstanding, LAB have a long history of being widely employed in the manufacturing of fermented foods and are usually considered safe. Some strains of LAB have been shown to have intrinsic or acquired antibiotic resistance. LAB are one of the sources of antibiotic-resistant genes spread in an environment and then transmitted to humans and animals via the food chain. They are described and could transfer to pathogenic bacteria. It raises the danger of food, crop, livestock, and aquaculture contamination, but it also has an impact on human and animal health. As a result, antibiotic resistance in LAB is currently a global problem. It is critical to identify the antibiotic resistance of LAB in various fermented foods.

Consequently, from the information described above, it could be illustrated that although many alternative antimicrobial agents have been developed, bacteria could adapt themselves for survival. Using an appropriate starter and an acceptable substrate for fermentation is one strategy for preventing the spread of microorganisms that cause antibiotic resistance. Additionally, probiotic qualification, antibacterial resistance genes composition, and possibly disease-causing virulence genes and toxicological data of LAB in different fermented foods are necessary to be evaluated. The development of strategies to increase antibiotic susceptibility is a critical study in order to safety of using LAB in food, humans, and animals in the long-term application.

REFERENCES

Adolfsson, O., Meydani, S. N., and Russell, R. M. (2004). Yogurt and gut function. *Am J Clin Nutr*, *80*(2), 245–256.

Ahmed, M. O., and Baptiste, K. E. (2018). Vancomycin-resistant enterococci: A review of antimicrobial resistance mechanisms and perspectives of human and animal health. *Microb Drug Resist*, *24*(5), 590–606.

Álvarez-Cisneros, Y. M., and Ponce-Alquicira, E. (2018). Antibiotic resistance in lactic acid bacteria. In *Antimicrobial Resistance-A Global Threat*. IntechOpen, pp. 53–73.

Aly, S. M., and Albutti, A. (2014). Antimicrobials use in aquaculture and their public health impact. *J Aquac Res Dev*, *5*(4), 1.

Amerikova, M., Pencheva El-Tibi, I., Maslarska, V., Bozhanov, S., and Tachkov, K. (2019). Antimicrobial activity, mechanism of action, and methods for stabilisation of defensins as new therapeutic agents. *Biotechnol Biotechnol Equip*, *33*(1), 671–682.

Amiri, S., Aghamirzaei, M., Mostashari, P., Sarbazi, M., Tizchang, S., and Madahi, H. (2021). The impact of biotechnology on dairy industry. In *Microbial Biotechnology in Food and Health*. Elsevier. pp. 53–79.

Ammor, M. S., Flórez, A. B., and Mayo, B. (2007). Antibiotic resistance in non-enterococcal lactic acid bacteria and bifidobacteria. *Food Microbiol*, *24*(6), 559–570.

Aristimuño Ficoseco, C., Mansilla, F. I., Maldonado, N. C., Miranda, H., Fátima Nader-Macias, M. E., and Vignolo, G. M. (2018). Safety and growth optimization of lactic acid bacteria isolated from feedlot cattle for probiotic formula design. *Front Microbiol*, *9*, 2220.

Bennedsen, M., Stuer-Lauridsen, B., Danielsen, M., and Johansen, E. (2011). Screening for antimicrobial resistance genes and virulence factors via genome sequencing. *Appl Environ Microbiol*, *77*(8), 2785–2787.

Biemer, J. J. (1973). Antimicrobial susceptibility testing by the Kirby-Bauer disc diffusion method. *Ann Clin Lab Sci*, *3*(2), 135–140.

Billström, H., Lund, B., Sullivan, Å., and Nord, C. E. (2008). Virulence and antimicrobial resistance in clinical *Enterococcus faecium*. *Int J Antimicrob Agents*, *32*(5), 374–377.

Binda, S., Hill, C., Johansen, E., Obis, D., Pot, B., Sanders, M. E., Tremblay, A., and Ouwehand, A. C. (2020). Criteria to qualify microorganisms as "probiotic" in foods and dietary supplements. *Front Microbiol*, *11*(1662), 1–9.

Binh, V. N., Dang, N., Anh, N. T. K., and Thai, P. K. (2018). Antibiotics in the aquatic environment of Vietnam: Sources, concentrations, risk and control strategy. *Chemosphere*, *197*, 438–450.

Bode, H. B. (2009). Entomopathogenic bacteria as a source of secondary metabolites. *Curr Opin Chem Biol*, *13*(2), 224–230.

Borak, M. D., Sarc, L., Mugerli, D. G., Antolic, B., and Brvar, M. (2020). Occupational inhalation poisoning with the veterinary antibiotic tiamulin. *Clin Toxicol*, *58*(4), 287–289.

Borges, S., Barbosa, J., Silva, J., and Teixeira, P. (2013). Evaluation of characteristics of *Pediococcus* spp. to be used as a vaginal probiotic. *J Appl Microbiol*, *115*(2), 527–538.

Bound, J. P., and Voulvoulis, N. (2005). Household disposal of pharmaceuticals as a pathway for aquatic contamination in the United Kingdom. *Environ Health Perspect*, *113*(12), 1705–1711.

Burridge, L., Weis, J. S., Cabello, F., Pizarro, J., and Bostick, K. (2010). Chemical use in salmon aquaculture: A review of current practices and possible environmental effects. *Aquaculture*, *306*(1–4), 7–23.

Busetti, A., Maggs, C. A., and Gilmore, B. F. (2017). Marine macroalgae and their associated microbiomes as a source of antimicrobial chemical diversity. *Eur J Phycol*, *52*(4), 452–465.

Cabello, F. C., Godfrey, H. P., Tomova, A., Ivanova, L., Dölz, H., Millanao, A., and Buschmann, A. H. (2013). Antimicrobial use in aquaculture re-examined: Its relevance to antimicrobial resistance and to animal and human health. *Environ Microbiol*, *15*(7), 1917–1942.

Calero-Cáceres, W., Melgarejo, A., Colomer-Lluch, M., Stoll, C., Lucena, F., Jofre, J., and Muniesa, M. (2014). Sludge as a potential important source of antibiotic resistance genes in both the bacterial and bacteriophage fractions. *Environ Sci Technol*, *48*(13), 7602–7611.

Cattoir, V., and Leclercq, R. (2017). Resistance to Macrolides, Lincosamides, and Streptogramins. In *Antimicrobial Drug Resistance*., Springer. pp. 269–280.

Cauwerts, K., Pasmans, F., Devriese, L., Martel, A., Haesebrouck, F., and Decostere, A. (2006). Cloacal *Lactobacillus* isolates from broilers show high prevalence of resistance towards macrolide and lincosamide antibiotics. *Avian Pathol*, *35*(02), 160–164.

Cavicchioli, V. Q., Camargo, A. C., Todorov, S. D., and Nero, L. A. (2019). Potential control of *Listeria monocytogenes* by bacteriocinogenic *Enterococcus hirae* ST57ACC and *Pediococcus pentosaceus* ST65ACC strains isolated from artisanal cheese. *Probiotics Antimicrob Proteins*, *11*(2), 696–704.

Chancey, S. T., Zähner, D., and Stephens, D. S. (2012). Acquired inducible antimicrobial resistance in Gram-positive bacteria. *Future Microbiol*, *7*(8), 959–978.

Chin, C.-S., Alexander, D. H., Marks, P., Klammer, A. A., Drake, J., Heiner, C., Clum, A., Copeland, A., Huddleston, J., Eichler, E. E., Turner, S. W., and Korlach, J. (2013). Nonhybrid, finished microbial genome assemblies from long-read SMRT sequencing data. *Nat Methods*, *10*(6), 563–569.

Clementi, F., and Aquilanti, L. (2011). Recent investigations and updated criteria for the assessment of antibiotic resistance in food lactic acid bacteria. *Anaerobe*, *17*(6), 394–398.

Collins, J., van Pijkeren, J. P., Svensson, L., Claesson, M. J., Sturme, M., Li, Y., Cooney, J. C., van Sinderen, D., Walker, A. W., and Parkhill, J. (2012). Fibrinogen-binding and platelet-aggregation activities of a *Lactobacillus salivarius* septicaemia isolate are mediated by a novel fibrinogen-binding protein. *Mol Microbiol*, *85*(5), 862–877.

Colombo, M., Castilho, N., Todorov, S. D., and Nero, L. A. (2018). Beneficial properties of lactic acid bacteria naturally present in dairy production. *BMC Microbiol*, *18*(1), 1–12.

Czaplewski, L., Bax, R., Clokie, M., Dawson, M., Fairhead, H., Fischetti, V. A., Foster, S., Gilmore, B. F., Hancock, R. E., and Harper, D. (2016). Alternatives to antibiotics – a pipeline portfolio review. *Lancet Infect Dis*, *16*(2), 239–251.

Damodharan, K., Lee, Y. S., Palaniyandi, S. A., Yang, S. H., and Suh, J.-W. (2015). Preliminary probiotic and technological characterization of *Pediococcus pentosaceus* strain KID7 and in vivo assessment of its cholesterol-lowering activity. *Front Microbiol*, *6*, 768.

Danielsen, M., and Wind, A. (2003). Susceptibility of *Lactobacillus* spp. to antimicrobial agents. *Int J Food Microbiol*, *82*(1), 1–11.

Danielsen, M., Simpson, P., O'connor, E., Ross, R., and Stanton, C. (2007). Susceptibility of *Pediococcus* spp. to antimicrobial agents. *J Appl Microbiol*, *102*(2), 384–389.

De Kwaadsteniet, M., Fraser, T., Van Reenen, C., and Dicks, L. (2006). Bacteriocin T8, a novel class IIa sec-dependent bacteriocin produced by *Enterococcus faecium* T8, isolated from vaginal secretions of children infected with human immunodeficiency virus. *Appl Environ Microbiol*, *72*(7), 4761–4766.

De Leener, E., Martel, A., Decostere, A., and Haesebrouck, F. (2004). Distribution of the erm (B) Gene, tet racycline Resistance Genes, and Tn 1545-like Transposons in Macrolide-and Lincosamide-Resistant Enterococci from Pigs and Humans. *Microb Drug Resist*, *10*(4), 341–345.

De Lima, C. J., Coelho, L. F., and Contiero, J. (2010). The use of response surface methodology in optimization of lactic acid production: Focus on medium supplementation, temperature and pH control. *Food Technol Biotechnol*, *48*(2), 175–181.

Dec, M., Urban-Chmiel, R., Stępień-Pyśniak, D., and Wernicki, A. (2017). Assessment of antibiotic susceptibility in *Lactobacillus* isolates from chickens. *Gut Pathog*, *9*(1), 1–16.

Diarra, M. S., Rempel, H., Champagne, J., Masson, L., Pritchard, J., and Topp, E. (2010). Distribution of antimicrobial resistance and virulence genes in *Enterococcus* spp. and characterization of isolates from broiler chickens. *Appl Environ Microbiol*, *76*(24), 8033–8043.

Dušková, M., and Karpíšková, R. (2013). Antimicrobial resistance of lactobacilli isolated from food. *Czech J Food Sci*, *31*(1), 27–32.

ECDC, EFSA, and EMA. (2017). ECDC/EFSA/EMA second joint report on the integrated analysis of the consumption of antimicrobial agents and occurrence of antimicrobial resistance in bacteria from humans and food-producing animals: Joint Interagency Antimicrobial Consumption and Resistance Analysis (JIACRA) Report. *EFSA J*, *15*(7), e04872.

EMA. (2014). Answers to the requests for scientific advice on the impact on public health and animal health of the use of antibiotics in animals. *European Medicines Agency*, *2014*, 381884.

Erginkaya, Z., Turhan, E., and Tatlı, D. (2018). Determination of antibiotic resistance of lactic acid bacteria isolated from traditional Turkish fermented dairy products. *Iran J Vet Res*, *19*(1), 53.

Esaiassen, E., Hjerde, E., Cavanagh, J. P., Simonsen, G. S., and Klingenberg, C. (2017). *Bifidobacterium* bacteremia: Clinical characteristics and a genomic approach to assess pathogenicity. *J Clin Microbiol*, *55*(7), 2234–2248.

European Commission (2003). Opinion of the scientific committee on animal nutrition on the criteria for assessing the safety of micro-organisms resistant to antibiotics of human clinical and veterinary importance. *Health & consumer protection directorate-general.* 1–21. Available: https://ec.europa.eu/food/fs/sc/scan/out108_en.pdf Accessed 2021 Nov 19

European Food Safety Authority (EFSA) (2012). Panel on additives products or substances used in animal feed, guidance on the assessment of bacterial susceptibility to antimicrobials of human and veterinary importance. *EFSA J*, *10*(6), 2740.

FAO/WHO (2002). *Guidelines for the Evaluation of Probiotics in Food.* Joint FAO/WHO Working Group Meeting, pp. 1–11.

Flórez, A. B., Delgado, S., and Mayo, B. (2005). Antimicrobial susceptibility of lactic acid bacteria isolated from a cheese environment. *Can J Microbiol*, *51*(1), 51–58.

Flórez, A. B., Ammor, M. S., Álvarez-Martín, P., Margolles, A., and Mayo, B. (2006). Molecular analysis of tet (W) gene-mediated tetracycline resistance in dominant intestinal *Bifidobacterium* species from healthy humans. *Appl Environ Microbiol*, *72*(11), 7377–7379.

Flórez, A. B., Vázquez, L., Rodríguez, J., and Mayo, B. (2021). Directed recovery and molecular characterization of antibiotic resistance plasmids from cheese bacteria. *Int J Mol Sci*, *22*(15), 7801.

Fouhy, F., O'Connell Motherway, M., Fitzgerald, G. F., Ross, R. P., Stanton, C., van Sinderen, D., and Cotter, P. D. (2013). In silico assigned resistance genes confer *Bifidobacterium* with partial resistance to aminoglycosides but not to β-lactams. *PLoS One*, *8*(12), e82653.

Fraqueza, M. J. (2015). Antibiotic resistance of lactic acid bacteria isolated from dry-fermented sausages. *Int J Food Microbiol*, *212*, 76–88.

Frese, S. A., Benson, A. K., Tannock, G. W., Loach, D. M., Kim, J., Zhang, M., Oh, P. L., Heng, N. C., Patil, P. B., and Juge, N. (2011). The evolution of host specialization in the vertebrate gut symbiont *Lactobacillus reuteri*. *PLoS Genet*, *7*(2), e1001314.

Gad, G. F. M., Abdel-Hamid, A. M., and Farag, Z. S. H. (2014). Antibiotic resistance in lactic acid bacteria isolated from some pharmaceutical and dairy products. *Braz J Microbiol*, *45*, 25–33.

Gaggìa, F., Mattarelli, P., and Biavati, B. (2010). Probiotics and prebiotics in animal feeding for safe food production. *Int J Food Microbiol*, *141*, S15–S28.

Gora, A. H., Rehman, S., Kiron, V., Dias, J., Fernandes, J.M., Olsvik, P. A., Siriyappagouder, P., Vatsos, I., Schmid-Staiger, U., Frick, K., and Cardoso, M. (2021). Management of Hypercholesterolemia Through Dietary ß-glucans–Insights From a Zebrafish Model. *Front Nutr*, *8*, 797452.

Graf, F. E., Palm, M., Warringer, J., and Farewell, A. (2019). Inhibiting conjugation as a tool in the fight against antibiotic resistance. *Drug Dev Res*, *80*(1), 19–23.

Griffith, M., Postelnick, M., and Scheetz, M. (2012). Antimicrobial stewardship programs: methods of operation and suggested outcomes. *Expert Rev Anti-Infect Ther*, *10*(1), 63–73.

Hanna, N., Sun, P., Sun, Q., Li, X., Yang, X., Ji, X., Zou, H., Ottoson, J., Nilsson, L. E., and Berglund, B. (2018). Presence of antibiotic residues in various environmental compartments of Shandong province in eastern China: its potential for resistance development and ecological and human risk. *Environ Int*, *114*, 131–142.

Harnisz, M. (2013). Total resistance of native bacteria as an indicator of changes in the water environment. *Environ Pollut*, *174*, 85–92.

Hashempour-Baltork, F., Hosseini, H., Shojaee-Aliabadi, S., Torbati, M., Alizadeh, A. M., and Alizadeh, M. (2019). Drug resistance and the prevention strategies in food borne bacteria: An update review. *Adv Pharm Bull*, *9*(3), 335.

Hollenbeck, B.L., and Rice, L.B. (2012). Intrinsic and acquired resistance mechanisms in enterococcus. *Virulence*, *3*(5), 421–569.

Hooper, D. C. (2001). Mechanisms of action of antimicrobials: Focus on fluoroquinolones. *Clin Infect Dis*, *32*(1), S9–S15.

Huang, M., Lu, J. J., and Ding, J. (2021). Natural products in cancer therapy: Past, present and future. *Nat prod bioprospect*, *11*(1), 5–13.

Hummel, A. S., Hertel, C., Holzapfel, W. H., and Franz, C. M. (2007). Antibiotic resistances of starter and probiotic strains of lactic acid bacteria. *Appl Environ Microbiol*, *73*(3), 730–739.

Ilavenil, S., Vijayakumar, M., Kim, D. H., Valan Arasu, M., Park, H. S., Ravikumar, S., and Choi, K. C. (2016). Assessment of probiotic, antifungal and cholesterol lowering properties of *Pediococcus pentosaceus* KCC-23 isolated from Italian ryegrass. *J Sci Food Agric*, *96*(2), 593–601.

Jacobsen, L., Wilcks, A., Hammer, K., Huys, G., Gevers, D., and Andersen, S. R. (2007). Horizontal transfer of tet (M) and erm (B) resistance plasmids from food strains of *Lactobacillus plantarum* to *Enterococcus faecalis* JH2-2 in the gastrointestinal tract of gnotobiotic rats. *FEMS Microbiol Ecol*, *59*(1), 158–166.

Jia, W., Li, G., and Wang, W. (2014). Prevalence and antimicrobial resistance of *Enterococcus* species: A hospital-based study in China. *Int J Environ Res Public Health*, *11*(3), 3424–3442.

Jung, D.-H., Chung, W.-H., Seo, D.-H., Nam, Y.-D., Yoon, S., and Park, C.-S. (2018). Complete genome sequence of *Bifidobacterium choerinum* FMB-1, a resistant starch-degrading bacterium. *J Biotechnol*, *274*, 28–32.

Jutkina, J., Marathe, N., Flach, C.-F., and Larsson, D. (2018). Antibiotics and common antibacterial biocides stimulate horizontal transfer of resistance at low concentrations. *Sci Total Environ*, *616*, 172–178.

Kim, M. J., Ku, S., Kim, S. Y., Lee, H. H., Jin, H., Kang, S., Li, R., Johnston, T. V., Park, M. S., and Ji, G. E. (2018). Safety evaluations of *Bifidobacterium bifidum* BGN4 and *Bifidobacterium longum* BORI. *Int J Mol Sci*, *19*(5), 1422.

Kimera, Z. I., Frumence, G., Mboera, L. E., Rweyemamu, M., Mshana, S. E., and Matee, M. I. (2020). Assessment of drivers of antimicrobial use and resistance in poultry and domestic pig farming in the Msimbazi river basin in Tanzania. *Antibiotics*, *9*(12), 838.

Kittinger, C., Kirschner, A., Lipp, M., Baumert, R., Mascher, F., Farnleitner, A. H., and Zarfel, G. E. (2018). Antibiotic resistance of *Acinetobacter* spp. isolates from the river Danube: Susceptibility stays high. *Int J Environ Res Public Health*, *15*(1), 52.

Kiwaki, M., and Sato, T. (2009). Antimicrobial susceptibility of *Bifidobacterium breve* strains and genetic analysis of streptomycin resistance of probiotic *B. breve* strain Yakult. *Int J Food Microbiol*, *134*(3), 211–215.

Klare, I., Konstabel, C., Werner, G., Huys, G., Vankerckhoven, V., Kahlmeter, G., Hildebrandt, B., Müller-Bertling, S., Witte, W., and Goossens, H. (2007). Antimicrobial susceptibilities of *Lactobacillus*, *Pediococcus* and *Lactococcus* human isolates and cultures intended for probiotic or nutritional use. *J Antimicrob Chemother*, *59*(5), 900–912.

Klein, E. Y., Van Boeckel, T. P., Martinez, E. M., Pant, S., Gandra, S., Levin, S. A., Goossens, H., and Laxminarayan, R. (2018). Global increase and geographic convergence in antibiotic consumption between 2000 and 2015. *Proc Natl Acad Sci*, *115*(15), E3463–E3470.

Klose, V., Bayer, K., Kern, C., Goelß, F., Fibi, S., and Wegl, G. (2014). Antibiotic resistances of intestinal lactobacilli isolated from wild boars. *Vet Microbiol*, *168*(1), 240–244.

Knapp, C. W., Dolfing, J., Ehlert, P. A., and Graham, D. W. (2010). Evidence of increasing antibiotic resistance gene abundances in archived soils since 1940. *Environ Sci Technol*, *44*(2), 580–587.

Koch, N., Islam, N. F., Sonowal, S., Prasad, R., and Sarma, H. (2021). Environmental antibiotics and resistance genes as emerging contaminants: Methods of detection and bioremediation. *Current Research in Microbial Sciences*, 2, 100027. https://doi.org/10.1016/j.crmicr.2021.100027

Korhonen, J. (2010). Antibiotic resistance of lactic acid bacteria. In*: Dissertations in Forestry and Natural Sciences*. Pasanen P., Lehto T., and Peiponen K. Kuopio, Eastern Finland. The University of Eastern Finland Dissertations in Forestry and Natural Sciences, pp. 15–72.

Krawczyk, B., Wityk, P., Gałęcka, M., and Michalik, M. (2021). The many faces of *Enterococcus* spp. – Commensal, probiotic and opportunistic pathogen. *Microorganisms*, *9*(9), 1900.

Kümmerer, K. (2009). Antibiotics in the aquatic environment–a review–part I. *Chemosphere*, *75*(4), 417–434.

Kushiro, A., Chervaux, C., Cools-Portier, S., Perony, A., Legrain-Raspaud, S., Obis, D., Onoue, M., and van de Moer, A. (2009). Antimicrobial susceptibility testing of lactic acid bacteria and bifidobacteria by broth microdilution method and Etest. *Int J Food Microbiol*, *132*(1), 54–58.

Ladha, G., and Jeevaratnam, K. (2018). Probiotic potential of *Pediococcus pentosaceus* LJR1, a bacteriocinogenic strain isolated from rumen liquor of goat (*Capra aegagrus* hircus). *Food Biotechnol*, *32*(1), 60–77.

Lakin, S. M., Dean, C., Noyes, N. R., Dettenwanger, A., Ross, A. S., Doster, E., Rovira, P., Abdo, Z., Jones, K. L., Ruiz, J., Belk, K. E., Morley, P. S., and Boucher, C. (2016). MEGARes: An antimicrobial resistance database for high throughput sequencing. *Nucleic Acids Res*, *45*(D1), D574–D580.

Larsson, D. J., de Pedro, C., and Paxeus, N. (2007). Effluent from drug manufactures contains extremely high levels of pharmaceuticals. *J Hazard Mater*, *148*(3), 751–755.

Laulund, S., Wind, A., Derkx, P. M., and Zuliani, V. (2017). Regulatory and safety requirements for food cultures. *Microorganisms*, *5*(2), 28.

Lima, L. M., da Silva, B. N. M., Barbosa, G., and Barreiro, E. J. (2020). β-lactam antibiotics: An overview from a medicinal chemistry perspective. *Eur J Med Chem*, *208*, 112829.

Lüdin, P., Roetschi, A., Wüthrich, D., Bruggmann, R., Berthoud, H., and Shani, N. (2018). Update on tetracycline susceptibility of *Pediococcus acidilactici* based on strains isolated from swiss cheese and whey. *J Food Prot*, *81*(10), 1582–1589.

Mahon, C., Lehman, D., and Manuselis, G. (2014). Antimicrobial agent mechanisms of action and resistance. *Int J Toxicol*, *22*, 135–143.

Mannu, L., Paba, A., Daga, E., Comunian, R., Zanetti, S., Duprè, I., and Sechi, L. A. (2003). Comparison of the incidence of virulence determinants and antibiotic resistance between *Enterococcus faecium* strains of dairy, animal and clinical origin. *Int J Food Microbiol*, *88*(2–3), 291–304.

Mathur, S., and Singh, R. (2005). Antibiotic resistance in food lactic acid bacteria – a review. *Int J Food Microbiol*, *105*(3), 281–295.

McManus, P. S., Stockwell, V. O., Sundin, G. W., and Jones, A. L. (2002). Antibiotic use in plant agriculture. *Annu Rev Phytopathol*, *40*(1), 443–465.

Missoum, A. (2018). DNA microarray and bioinformatics technologies: A mini-review. *Proc Royal Soc*, *2*(1), 02010.

Mitchell, C., Fredricks, D., Agnew, K., and Hitti, J. (2015). Hydrogen-peroxide producing lactobacilli are associated with lower levels of vaginal IL1β, independent of bacterial vaginosis. *Sex Transm Dis*, *42*(7), 358.

Mohanty, S., and Bhuniya, S. (2021). Pleural empyema due to *Enterococcus* species: An uncommon etiology. *Respir Med Case Rep*, *32*, 101375.

Morar, A., Sala, C., and Imre, K. (2015). Occurrence and antimicrobial susceptibility of *Salmonella* isolates recovered from the pig slaughter process in Romania. *J Infect Dev Ctries*, *9*(1), 099–104.

Muñoz-Atienza, E., Gómez-Sala, B., Araújo, C., Campanero, C., Del Campo, R., Hernández, P. E., Herranz, C., and Cintas, L. M. (2013). Antimicrobial activity, antibiotic susceptibility and virulence factors of lactic acid bacteria of aquatic origin intended for use as probiotics in aquaculture. *BMC Microbiol*, *13*(1), 1–22.

Nayyar, S., Kadyan, S., Parveen, H., Mallappa, R. H., and Pradhan, D. (2022). Antimicrobial resistance in food grade lactic acid bacteria. In *Biological and Chemical Hazards in Food and Food Products*. Apple Academic Press. pp. 91–120.

Ouoba, L. I. I., Lei, V., and Jensen, L. B. (2008). Resistance of potential probiotic lactic acid bacteria and bifidobacteria of African and European origin to antimicrobials: determination and transferability of the resistance genes to other bacteria. *Int J Food Microbiol*, *121*(2), 217–224.

Ouwehand, A., and Röytiö, H. (2015). Probiotic fermented foods and health promotion. In: *Advances in Fermented Foods and Beverages*. Woodhead Publishing, pp. 3–22.

Padmasini, E., Padmaraj, R., and Ramesh, S. S. (2014). High level aminoglycoside resistance and distribution of aminoglycoside resistant genes among clinical isolates of *Enterococcus* species in Chennai, India. *Sci World J*, *2014*, 329157.

Patel, A., Shah, N., and Prajapati, J. (2012). Antibiotic resistance profile of lactic acid bacteria and their implications in food chain. *World J Dairy Food Sci*, *7*(2), 202–211.

Preethi, C., Thumu, S. C. R., and Halami, P. M. (2017). Occurrence and distribution of multiple antibiotic-resistant *Enterococcus* and *Lactobacillus* spp. from Indian poultry: In vivo transferability of their erythromycin, tetracycline and vancomycin resistance. *Ann Microbiol*, *67*(6), 395–404.

Pruden, A., Larsson, D. J., Amézquita, A., Collignon, P., Brandt, K. K., Graham, D. W., Lazorchak, J. M., Suzuki, S., Silley, P., and Snape, J. R. (2013). Management options for reducing the release of antibiotics and antibiotic resistance genes to the environment. *Environ Health Perspect*, *121*(8), 878–885.

Rabello, R. F., Bonelli, R. R., Penna, B. A., Albuquerque, J. P., Souza, R. M., and Cerqueira, A. M. (2020). Antimicrobial resistance in farm animals in Brazil: An update overview. *Animals*, *10*(4), 552.

Raeisi, J., Saifi, M., Pourshafie, M. R., Habibi, M., Mohajerani, H. R., Akbari, N., and Karam, M. R. A. (2017). Rapid identification of vancomycin resistant *Enterococcus faecalis* clinical isolates using a sugar fermentation method. *JCDR*, *11*(3), DC14.

Rakhmanova, A., Khan, Z., and Shah, K. (2018). A mini review fermentation and preservation: Role of lactic acid bacteria. *MOJFPT*, *6*(5), 414–417.

Raymann, K., Shaffer, Z., and Moran, N. A. (2017). Antibiotic exposure perturbs the gut microbiota and elevates mortality in honeybees. *PLoS Biol*, *15*(3), e2001861.

Reygaert, W. C. (2018). An overview of the antimicrobial resistance mechanisms of bacteria. *AIMS Microbiol*, *4*(3), 482.

Rice, L. B. (2012). Mechanisms of resistance and clinical relevance of resistance to β-lactams, glycopeptides, and fluoroquinolones. *Mayo Clin. Proc*, *87*(2), 198–208.

Rodriguez-Alonso, P., Fernandez-Otero, C., Centeno, J., and Garabal, J. (2009). Antibiotic resistance in lactic acid bacteria and Micrococcaceae/Staphylococcaceae isolates from artisanal raw milk cheeses, and potential implications on cheese making. *J Food Sci*, *74*(6), M284–M293.

Rossi, F., Amadoro, C., and Colavita, G. (2019). Members of the *Lactobacillus* genus complex (LGC) as opportunistic pathogens: A review. *Microorganisms*, *7*(5), 126.

Rozman, V., Lorbeg, P. M., Accetto, T., and Matijašić, B. B. (2020). Characterization of antimicrobial resistance in lactobacilli and bifidobacteria used as probiotics or starter cultures based on integration of phenotypic and in silico data. *Int J Food Microbiol*, *314*, 108388.

Ruas-Madiedo, P., Gueimonde, M., Fernández-García, M., de los Reyes-Gavilán, C. G., and Margolles, A. (2008). Mucin degradation by *Bifidobacterium* strains isolated from the human intestinal microbiota. *Appl Environ Microbiol*, *74*(6), 1936–1940.

Rugumisa, B., Call, D., Mwanyika, G., Subbiah, M., and Buza, J. (2016). Comparison of the prevalence of antibiotic-resistant *Escherichia coli* isolates from commercial-layer and free-range chickens in Arusha district, Tanzania. *Afr J Microbiol Res*, *10*(34),1422–1429.

Rusch, M., Spielmeyer, A., Zorn, H., and Hamscher, G. (2019). Degradation and transformation of fluoroquinolones by microorganisms with special emphasis on ciprofloxacin. *Appl Microbiol Biotechnol*, *103*(17), 6933–6948.

Salvetti, E., and O'Toole, P. W. (2017). When regulation challenges innovation: The case of the genus *Lactobacillus*. *Trends Food Sci Technol*, *66*, 187–194.

Schar, D., Klein, E. Y., Laxminarayan, R., Gilbert, M., and Van Boeckel, T. P. (2020). Global trends in antimicrobial use in aquaculture. *Sci Rep*, *10*(1), 1–9.

Seal, B. S., Drider, D., Oakley, B. B., Brüssow, H., Bikard, D., Rich, J. O., Miller, S., Devillard, E., Kwan, J., and Bertin, G. (2018). Microbial-derived products as potential new antimicrobials. *Vet Res*, *49*(1), 1–12.

Serwecińska, L. (2020). Antimicrobials and antibiotic-resistant bacteria: A risk to the environment and to public health. *Water*, *12*(12), 3313.

Shazali, N., Foo, H. L., Loh, T. C., Choe, D. W., and Rahim, R. A. (2014). Prevalence of antibiotic resistance in lactic acid bacteria isolated from the faeces of broiler chicken in Malaysia. *Gut Pathogens*, *6*(1), 1–7.

Shukla, R., and Goyal, A. (2014). Probiotic potential of *Pediococcus pentosaceus* CRAG3: A new isolate from fermented cucumber. *Probiotics Antimicrob Proteins*, *6*(1), 11–21.

Sirichoat, A., Flórez, A. B., Vázquez, L., Buppasiri, P., Panya, M., Lulitanond, V., and Mayo, B. (2020). Antibiotic susceptibility profiles of lactic acid bacteria from the human vagina and genetic basis of acquired resistances. *Int J Mol Sci*, *21*(7), 2594.

Soge, O. O., Beck, N. K., White, T. M., No, D. B., and Roberts, M. C. (2008). A novel transposon, Tn 6009, composed of a Tn 916 element linked with a *Staphylococcus aureus* mer operon. *J Antimicrob Chemother*, *62*(4), 674–680.

Stefańska, I., Kwiecień, E., Jóźwiak-Piasecka, K., Garbowska, M., Binek, M., and Rzewuska, M. (2021). Antimicrobial susceptibility of lactic acid bacteria strains of potential use as feed additives-the basic safety and usefulness criterion. *Front Vet Sci*, *8*, 1–11.

Sundin, G. W., and Wang, N. (2018). Antibiotic resistance in plant-pathogenic bacteria. *Annu Rev Phytopathol*, *56*, 161–180.

Tello, A., Austin, B., and Telfer, T. C. (2012). Selective pressure of antibiotic pollution on bacteria of importance to public health. *Environ Health Perspect*, *120*(8), 1100–1106.

Temmerman, R., Pot, B., Huys, G., and Swings, J. (2003). Identification and antibiotic susceptibility of bacterial isolates from probiotic products. *Int J Food Microbiol*, *81*(1), 1–10.

Thiele-Brun, S., and Peters, D. (2007). Photodegradation of pharmaceutical antibiotics on slurry and soil surfaces. *Landbauforschung Volkenrode*, *57*(1), 13.

Thomas, K. V., Dye, C., Schlabach, M., and Langford, K. H. (2007). Source to sink tracking of selected human pharmaceuticals from two Oslo city hospitals and a wastewater treatment works. *Environ Monit Assess*, *9*(12), 1410–1418.

Tiseo, K., Huber, L., Gilbert, M., Robinson, T. P., and Van Boeckel, T. P. (2020). Global trends in antimicrobial use in food animals from 2017 to 2030. *Antibiotics*, *9*(12), 918.

Tong, D. Y. (2021). Antibiotic resistance in lactic acid bacteria. *J probiotics health*, *9*(2), 1–4.

Toomey, N., Bolton, D., and Fanning, S. (2010). Characterisation and transferability of antibiotic resistance genes from lactic acid bacteria isolated from Irish pork and beef abattoirs. *Res Microbiol*, *161*(2), 127–135.

Urbaniak, M., Wyrwicka, A., Tołoczko, W., Serwecińska, L., and Zieliński, M. (2017). The effect of sewage sludge application on soil properties and willow (*Salix* sp.) cultivation. *Sci Total Environ*, *586*, 66–75.

Van Boeckel, T. P., Brower, C., Gilbert, M., Grenfell, B. T., Levin, S. A., Robinson, T. P., Teillant, A., and Laxminarayan, R. (2015). Global trends in antimicrobial use in food animals. *Proc Natl Acad Sci*, *112*(18), 5649–5654.

Verraes, C., Van Boxstael, S., Van Meervenne, E., Van Coillie, E., Butaye, P., Catry, B., De Schaetzen, M.-A., Van Huffel, X., Imberechts, H., and Dierick, K. (2013). Antimicrobial resistance in the food chain: A review. *Int J Environ Res Public Health*, *10*(7), 2643–2669.

Victor, L. Y. (2011). Guidelines for hospital-acquired pneumonia and health-care-associated pneumonia: A vulnerability, a pitfall, and a fatal flaw. *Lancet Infect Dis*, *11*(3), 248–252.

Vieco-Saiz, N., Belguesmia, Y., Raspoet, R., Auclair, E., Gancel, F., Kempf, I., and Drider, D. (2019). Benefits and inputs from lactic acid bacteria and their bacteriocins as alternatives to antibiotic growth promoters during food-animal production. *Front Microbiol*, *10*, 57.

Vignesh, S., Dahms, H.-U., Muthukumar, K., Vignesh, G., and James, R. A. (2016). Biomonitoring along the tropical southern Indian coast with multiple biomarkers. *PLoS One*, *11*(12), e0154105.

Walters, E., McClellan, K., and Halden, R. U. (2010). Occurrence and loss over three years of 72 pharmaceuticals and personal care products from biosolids–soil mixtures in outdoor mesocosms. *Water Res*, *44*(20), 6011–6020.

Walther, C., Rossano, A., Thomann, A., and Perreten, V. (2008). Antibiotic resistance in *Lactococcus* species from bovine milk: Presence of a mutated multidrug transporter mdt (A) gene in susceptible *Lactococcus garvieae* strains. *Vet Microbiol*, *131*(3–4), 348–357.

Wang, J., Zhao, Z., Chen, J., Lu, H., Liu, G., Zhou, J., and Guan, X. (2017). PAHs accelerate the propagation of antibiotic resistance genes in coastal water microbial community. *Environ Pollut*, *231*, 1145–1152.

Wang, K., Conlon, M., Ren, W., Chen, B. B., and Bączek, T. (2018). Natural products as targeted modulators of the immune system. *J Immunol Res*, *2018*, 7862782.

Wendlandt, S., Lozano, C., Kadlec, K., Gómez-Sanz, E., Zarazaga, M., Torres, C., and Schwarz, S. (2013). The enterococcal ABC transporter gene *lsa* (E) confers combined resistance to lincosamides, pleuromutilins and streptogramin A antibiotics in methicillin-susceptible and methicillin-resistant *Staphylococcus aureus*. *J Antimicrob Chemother*, *68*(2), 473–475.

WHO. (2011). *Critically Important Antimicrobials for Human Medicine*. Geneva. Switzerland: WHO.

Wright, A. V. (2005). Regulating the safety of probiotics-the European approach. *Curr Pharm Des*, *11*(1), 17–23.

Yang, Y., Chen, D., and Huang, M. (2010). The source of antibiotics in the environment and progress of its ecological impact research. *Environ Sci Manag*, *35*(1), 140–143.

Yoon, Y., Dodd, M. C., and Lee, Y. (2018). Elimination of transforming activity and gene degradation during UV and UV/H_2O_2 treatment of plasmid-encoded antibiotic resistance genes. *Environ Sci Water Res Technol*, *4*(9), 1239–1251.

Zhao, Y., Yang, Q. E., Zhou, X., Wang, F. H., Muurinen, J., Virta, M. P., Brandt, K. K., and Zhu, Y. G. (2021). Antibiotic resistome in the livestock and aquaculture industries: Status and solutions. *Crit Rev Environ Sci Technol*, *51*(19), 2159–2196.

Zheng, M., Zhang, R., Tian, X., Zhou, X., Pan, X., and Wong, A. (2017). Assessing the risk of probiotic dietary supplements in the context of antibiotic resistance. *Front Microbiol*, *8*, 908.

Zhou, N., Zhang, J., Fan, M., Wang, J., Guo, G., and Wei, X. (2012). Antibiotic resistance of lactic acid bacteria isolated from Chinese yogurts. *Int J Dairy Sci*, *95*(9), 4775–4783.

Zhou, Y., Jiang, T., Hu, S., Wang, M., Ming, D., and Chen, S. (2017). Genomic insights of *Pannonibacter phragmitetus* strain 31801 isolated from a patient with a liver abscess. *Microbiologyopen*, *6*(6), e00515.

Zommiti, M., Bouffartigues, E., Maillot, O., Barreau, M., Szunerits, S., Sebei, K., Feuilloley, M., Connil, N., and Ferchichi, M. (2018). In vitro assessment of the probiotic properties and bacteriocinogenic potential of *Pediococcus pentosaceus* MZF16 isolated from artisanal Tunisian meat "Dried Ossban". *Front Microbiol*, *9*, 2607.

2 Antimicrobial Indexing of Indigenous Pathogens from Sanitary Landfill Leachate

Rachna Jain and Pritam Dey

2.1 INTRODUCTION

The discovery of antibiotics was one of the most serendipitous events in the history of science; they were developed for the treatment of different deadly bacterial infections, which used to take the lives of hundreds and thousands of people in little time. The first commercial antibiotic *penicillin* was developed by Alexander Fleming in 1928. In his Nobel Prize acceptance speech itself, he expressed his fear of the development of antibiotic-resistant bacteria. He quoted, "Then there is the danger that the ignorant man may easily underdose himself and, by exposing his microbes to non-lethal quantities of the drug, make them resistant."

His fear has come true after so many decades with an exponential increase in antibiotic consumption in "India, China, and USA." These countries reported the highest per capita antibiotic consumption rate for human therapeutics in 2010 (Boeckel et al., 2014).

A major proportion of this exponential rise lies in the heedless and unnecessary consumption of antibiotics, even for non-bacterial infections like the common cold, etc., which don't respond to antibiotics (Valitalo et al., 2017). Residual and unused/expired antibiotics are released into the environment and pooled down either in the soil matrix or in the wastewater/leachate pool. Continuous exposure of these antibiotics to native microflora leads to the development of resistance by different mechanisms, i.e., horizontal gene transfer, plasmid-based gene transfer, etc. Antimicrobial resistance (AMR) has been listed as a top threat in a recent report released by WHO in 2019. The abundance of AMR and Antibiotic resistance genes (ARGs) in wastewater leads to the development of a new surveillance tool, wastewater-based epidemiology (WBE). It monitors the presence of AMRs and ARGs in different wastewater systems and thus can help in the prediction of any disease outbreak leading to preventing its transmission.

Existing wastewater/sewage/landfill leachate treatment plants are not equipped to block the influx of antibiotics or medicinal residues. Also, generally, they don't use any specific treatment technology for the removal of residual antibiotics, which leads to the spread of antimicrobial resistance against frequently used drugs ciprofloxacin, erythromycin, and tetracycline (da Silva et al., 2006). It is evaluated that more or less 80% of wastewater consisting of millions of ARGs is discharged directly into the environment without sufficient treatment (Sekizuka et al., 2018). Al-Gheethi et al.(2018) have also stated that applied wastewater treatment technology reduced the total microbial load in the effluent but eventually increased the load of single and multi-drug resistant bacteria and ARGs in the effluent as well as settled sludge.

DOI: 10.1201/9781003268895-2

2.1.1 Classification of Antibiotics

Antibiotics are chemically bonded, cyclic structures which are developed to combat diverse bacterial infections. Based on their mechanism of action, they are classified into seven different classes, as detailed in Table S1 provided in the supplementary sheet.

2.1.2 Mechanisms of Antibiotic Resistance

2.1.2.1 Origin of Resistance

Bacterial sensitivity towards antimicrobial agents lacks uniformity and varies between closely related species. It is measured in the form of the minimum amount of antibiotic required to reduce the growth of bacteria and is called Minimum Inhibitory Concentration (MIC) (Kowalska-Krochmal et al., 2021). Resistance or susceptibility of bacteria depends upon average MIC; if it is higher, the bacterial species are termed resistant, and if found lower, susceptible bacteria dominate. Along with this, bacterial isolates may exhibit resistance through gene transfer from related species; the level of resistance varies with the type of species and characteristic of the acquired gene (Reygaert, 2018). Details are given in Tables S2 and S3 provided in the supplementary sheet.

On the basis of this, bacterial resistance may be classified into two parts:

1.1.2.1.1 Natural Resistance

It is the inherent metabolic property of bacteria present due to dominant expressive genes that show resistance towards different antimicrobial agents or can be induced by exposure to specific antimicrobial agents.

1.1.2.1.2 Acquired Resistance

It is a type of resistance acquired by bacterial species during their life cycle through different genetic routes. This may be temporary or permanently based on the type of mechanism as detailed below:

- Plasmid-mediated transmission
- Bacteriophage-borne transmission
- Mutation- Many such mutations hinder microbial growth; methicillin resistance in *Staphylococcus aureus* is acquired at the cost of decreased growth rate (Becker et al., 2018).
- By means of UV radiation or exposure to toxic chemicals (Reygaert, 2018).
- Through Insertion sequences and Integrins- substitution or deletion mutation in certain specific genes encoding drug receptors, drug transporter, and enzymes that cause antibiotic modifications (Blázquez et al., 2012).

The most challenging part of antimicrobial resistance is that it can be acquired even by consuming sub-lethal concentrations of common antibiotics on a regular basis. Acquired resistance passes onto different generations of hypermutable strains, may accelerate the acquisition of resistance to other related antimicrobials, and even can promote transposon movement (Martinez, 2014).

2.1.3 Antibiotic Resistance Genes (ARGs) and Their Environmental Dissemination

Apart from antimicrobial resistance bacteria, ARGs also possess significant environmental and human health impacts. These work like inert vectors that can be easily inserted into a native microbial population; their transfer mechanism into native pathogenic as well as non-pathogenic microbial species is depicted in Figure 2.1.

In view of these facts, we can state that antibiotics and antibiotic-resistant bacteria are omnipresent in different environmental matrices and have become a potential environmental and human

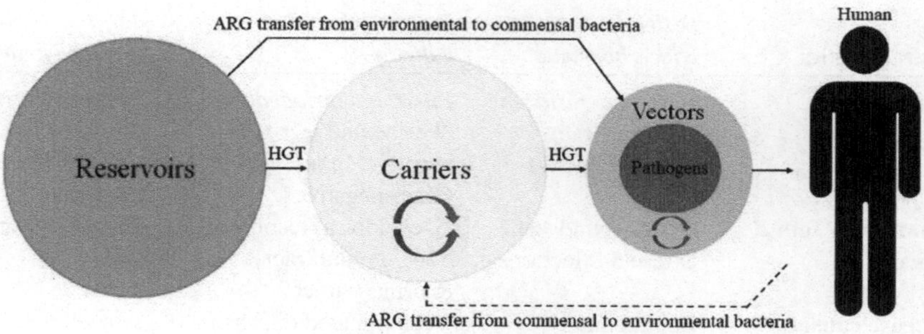

FIGURE 2.1 Dissemination of ARGs in the environment.

health threat. Therefore, it is necessary to monitor their presence in different wastewater bodies on a regular basis, apart from the normal physicochemical parameters. The present book chapter deals with AMR indexing of indigenous pathogenic microbes from sanitary landfill leachate to estimate their potential health and environmental impacts.

2.2 MATERIALS AND METHODOLOGY

2.2.1 Antibiotics

High-purity antibiotics were used for experiments as detailed below:

Sl No	Class of antibiotics	Antibiotic discs concentration	Abbreviation	HiMedia catalogue no.s
1.	Aminoglycosides	Vancomycin 30 µg	VA	SD045
		Gentamicin 50 µg	GEN	SD166
2.	Carbipenem	Imipenem 10 mcg	IPM	SD073
3.	Cephalosporins	Ceftadiazime 10 mcg	CAZ	SD062A
		Ceftriaxone 30 mcg	CZX	SD110
		Cefotaxime 30 mcg	CTX	SD040
		Cefixime 5 µg	CFM	SD211
4.	Cyclins	Oxytetracycline 30 µg	O	SD027
		Tetracycline 30 µg	TE	SD037
5.	Fluoroquinolones	Ofloxacin 5 µg	OF	SD087
6.	Macrolide	Erythromycin 10 µg	E	SD083
7.	Quinolones	Ciprofloxacin 10 µg	CIP	SD080
8.	β-Lactam	Ampicillin 10 µg	AMP	SD002
		Penicillin 10 µg/2 units	P	SD144
		Amoxicillin/ Clavulanic acid 20/10 µg	AMC	SD063
		Chloramphenicol 25 µg	C	SD153

2.2.2 Micro-organisms

Pure cultures of *Leclercia adecarboxylata*, *Pseudescherichia vulneris*, *Brucella inopinata* previously isolated from sanitary landfill leachate were used for this study. The morphological and metabolic characteristics of isolated strains are as follows:

Sl. No	Characteristics	*Leclercia adecarboxylata*	*Pseudoescherichia vulneris*	*Brucella inopinata*
1.	**Family**	*Enterobacteriaceae*	*Enterobacteriaceae*	*Enterobacteriaceae*
2.	**Morphology**	Motile, Gram-negative bacillus	"Fermentative, oxidase-negative, motile rod, Gram-negative"	Gram-negative, facultative, intracellular bacteria
3.	**Commonly found places**	Open wounds and external infections	"Develops at respiratory tract, genital tract, stool, & urinary tract."	Breast implant infection site
4.	**Disease caused**	Bacteremia	"Complicated diarrhea and sepsis in an infant."	Brucellosis
5.	**Antibiotic sensitivity**	Rare antibiotic resistance was observed for first-generation cephalosporin and SMX-TMP.	"Susceptible to 14 antibiotics, including third-generation cephalosporins, aminoglycosides, trimethoprim, and sulfamethoxazole-trimethoprim. Similar studies have shown they have some type of resistance to the antibiotics penicillin and clindamycin and were also marginally resistant to carbenicillin, erythromycin, tetracycline, chloramphenicol, and nitrofurantoin."	Susceptible to doxycycline, streptomycin, rifampin, trimethoprim-sulfamethoxazole low MIC against Tigecycline

2.2.3 Microbial Inoculum Preparation

Individual pure culture of *Leclercia* spp., *Pseudoescherichia* spp., and *Brucella* spp. was freshly made onto minimal agar media slant and overnight incubated at 37°C. The liquid culture was then made by inoculating individual isolates from fresh agar slants into a minimal broth medium and incubated at the same conditions to attain sufficient metabolic growth (0.5 McFarland unit) for approximately 18–24 hrs in an incubator shaker at 37°C and 100 rpm.

2.2.4 Experimental Methodology

Antibiotic susceptibility of indigenous isolates was identified in the form of Multi-Drug Resistance (MDR) profile and Multiple Antibiotic Resistance (MAR) Index:

2.2.4.1 Multi-drug Resistance (MDR)

Multi-drug resistance is an AMR profile of individual microbial culture that gives the broadways glimpse of microbial response towards a range of antibiotics. Overnight grown pure culture 100 µl has been inoculated individually onto minimal agar medium by standard spread plate method. Antibiotic sensitivity was tested against different listed antibiotics by the standard disc diffusion method. Plates were incubated at 37°C for 24 hrs. Results were depicted as the presence and absence of halo zone formation. Halo zone diameter against each susceptible antibiotic was measured in millimeters (mm).

2.2.4.2 Multiple Antibiotic Resistance (MAR) Index

It is the health risk index generated due to exposure to multiple antibiotics and can be determined by using the formula:

$$MAR = a/b,$$

a = Number of antibiotics to which the test isolate depicted resistance
b = Total Number of tested antibiotics (Sivaranjani, 2013).

Appropriate microbial controls were made and incubated under the same experimental conditions for validation.

2.3 RESULTS AND DISCUSSION

2.3.1 MDR Profile

Multiple Drug Resistance (MDR) is a bacterial phenotype showing resistance to ≥2 antimicrobial agents. All three indigenous isolates showed varied responses to tested antibiotics, as listed below in Table 2.1. *Pseudoescherichia vulneris* was found most susceptible to tested antibiotics and showed resistance to only two of the 16 tested. *Leclercia adecarboxylata* followed next, showing resistance to five antibiotics. *Brucella inopinata* was observed to be the most resistant microbes and showed resistance to 50% of the experimental antibiotics. The average halo zone diameter of all three isolates followed the order *Brucella inopinata* (0.95 mm) <*Pseudoescherichia vulneris* (1 mm) < *Leclercia adecarboxylata* (1.3 mm), as shown in Figure 2.2. This confirms *Brucella inopinata* as the most resistant microbial strain, but in the case of *Pseudoescherichia vulneris* and *Leclercia adecarboxylata*the order reversed; the higher average zone diameter of *Leclericia* spp. towards susceptible antibiotics, whereas *Pseusoescherichia* spp. showed a lower average diameter. It signifies

TABLE 2.1
MDR Profile of Isolated Strains

	Tested Antibiotics \ Microbial Strains		Halo Zone Diameter (mm)		
Sl No	**Class of antibiotics**	**Abbreviation**	***Leclercia adecarboxylata***	***Pseudoescherichia vulneris***	***Brucella inopinata***
1.	Aminoglycosides	VA	0.53±0.15	0.8±0.1	R
		GEN	1.1±0.1	1.1±0.17	1.37±0.15
2.	Carbipenem	IPM	1.37±0.15	1±0.1	1.37±0.15
3.	Cephalosporins	CAZ	R	1.37±0.15	R
		CZX	1.37±.015	1.13±.015	1.03±.021
		CTX	1.17±.015	1.17±0.15	1.1±0.2
		CFM	1.1±0.1	1.27±0.25	R
4.	Cyclins	O	1.5±0.3	1.33±0.15	0.67±0.15
		TE	2.23±0.25	0.33±0.15	0.53±0.06
5.	Fluoroquinolones	OF	1.97±0.15	R	0.63±0.15
6.	Macrolide	E	R	0.37±.015	R
7.	Quinolones	CIP	1.8±0.2	0.93±0.21	R
8.	β-Lactam	AMP	0.87±0.15	0.77±0.15	R
		P	R	R	0.63±0.15
		AMC	R	1.13±0.15	R
		C	R	1.1±0.1	R

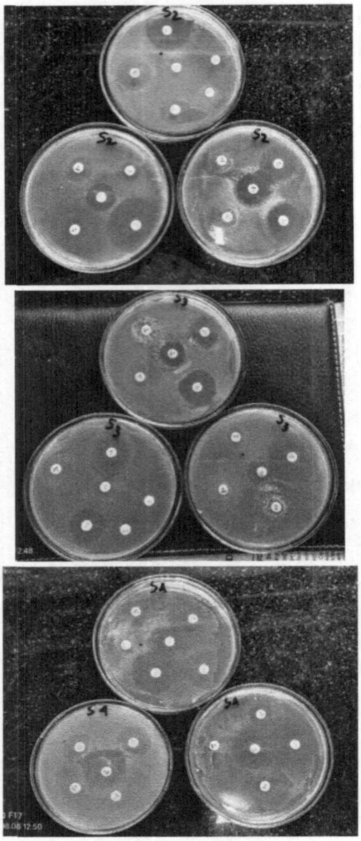

FIGURE 2.2 MDR profile of isolated strains.

that although *Pseusoescherichia* spp. is resistant to only two tested antibiotics, its susceptibility is not very prominent to the other 14 tested antibiotics.

Varied antimicrobial responses to tested antibiotics can be attributed to the presence of these antibiotics in landfill leachate. β lactams are commonly used high-consumption antibiotics used in India; hence higher resistance has been found for ampicillin, penicillin, tetracyclin, erythromycin, etc., whereas selective antibiotics, viz. vancomycin, amoxycillin, cefixime, and ciprofloxacin are less consumed; hence micro-organisms exhibit less resistance towards them. Similar significant associations between sampling locations and relative antimicrobial resistance in indigenous isolates were observed by Mukherjee et al., 2021. This ropes WWTP effluents as a source of disseminating antibiotic-resistant bacteria. In our study also, 75% of isolates showed resistance against ampicillin, erythromycin, and ceftadiazime.

A prevalent association of tetracycline-resistant bacteria in agricultural and animal husbandry effluent has been previously reported by Scaria et al. (2021). Similarly, a>90% resistance rate was observed in *E. coli* isolated from effluent collected from a swine lagoon (Brooks & McLaughlin, 2009). In addition, abundant tetracycline-resistant bacteria, as well as genes had been reported in different water samples collected from a watershed. Imipenem belongs to the carbapenem family and is used as the last line of defense against Gram-negative plant pathogenic microbes and is rarely used for human consumption (Nicolau et al., 2012). In the present study, all three bacterial isolates were found susceptible to Imipenem.

Higher resistance rates are mostly due to conjugative plasmid DNA transfer, as observed among *E.coli* isolates at different stages of the WWTP plant (Amador et al., 2015). As per the conventional definition, multi-drug resistance refers to resistance against two or more antimicrobial agents, but now it has been extended to "resistance to three or more classes" to define multi-drug resistant

Gram-positive and Gram-negative bacteria (Magiorakos et al., 2012). In this study, two of the three tested strains, *Brucella* and *Leclercia* showed resistance to more than two classes of antibiotics. The results are allied to other reports of the prevalence of MDR in *E. coli* from different surface water samples (Blaak et al., 2015), although they vary as a function of the tested antimicrobial agent as well as the site of sample collection (Figure 2.2).

2.3.2 Multiple Antibiotic Resistance (MAR) Index

The high prevalence of multi-drug resistance among the three bacterial isolates indicates a vital need for broad-based, local antimicrobial-sensitive surveillance and planning of effective interventions to decrease multi-drug resistance in such pathogens. The MAR index is a common tool for analyzing health risks and has been very helpful in checking the spread of bacterial resistance in a given population where there is resistance to more than three antibiotics (Mukherjee et al., 2021).

Accordingly, the MAR index in the present study was observed to be in the range of 0.1–0.5, the highest for *Brucella* spp. (0.5) followed by *Leclercia* spp. (0.3125) and lowest for *Pseudoescherichia* spp. (0.125). Hence, all three isolates fall under mild to medium health risk. Multiple antibiotic

TABLE 2.2
Review of Literature on Antimicrobial Resistance and ARGs in Landfill Leachate

Types of investigated samples	Location	Sample characteristics	Target Antibiotics	Key Findings	References
Landfill Leachate	Kumasi, Ghana	High antibiotic concentration has been observed in leachate	"Penicillin Metronidazole Amoxicillin"	"Indigenous isolates, *Enterobacteriaceae*, and some *Bacillus* and *Listeria* were found to be resistant."	Borquaye et al., 2019
Landfill Leachate	12 different landfill sites in China	Low COD and BOD_5 pertains to an old landfill site		"Five ARGs- tetO, tetW, blaTEM, sulI, and sulII and seven heterotrophic ARBs were extensively detected in all samples."	Wang et al., 2015
Landfill Leachate	Foz do Iguaçu City 25°27'47.9"S 54°36'26.4"W, in western Paraná State	Dark in color	"Amoxicillin, Potassium Clavulanate, Azithromycin and Chloramphenicol	"6 bacterial showed resistance to at least one tested antibiotic, while six fungal isolates showed resistance to the antimycotic nystatin."	Garcete et al., 2022
Landfill Leachate	13 different landfills located in different geographical locations of China		22 different Antibiotics	"526 ARG subtypes belonging to 21 ARG types were detected, with abundances ranging from	Zhao et al., 2018
Landfill Leachate	7 different landfills from 6 provinces of China	High Antibiotic load	" OFL, NOR, ENR, PEF, and the target BLs included CEF and AMOX"	"Targeted ARGs subtypes 1.44 ± 4.64×10−4 ARGs/16S and MGEs 7.88×10−2 ± 1.18×10−1 ARGs/16S were the highest."	You et al., 2018

resistance MAR in bacteria is very commonly related to the presence of plasmids with one or more resistance genes, each encoding a single antibiotic-sensitive phenotype. In concurrence with our study Christopher et al. (2013) also reported low to mild health risk assessment for various *E. coli* strains isolated from clinical specimens. MAR index varied from 0.18–0.89, with a majority beyond 0.6.

Some previously done studies related to the presence of AMR in leachate have been compiled below for a better understanding of this potentially threatening issue (Table 2.2).

2.4 RECOMMENDATIONS AND MEASURES TO BE TAKEN

The following reported strategies can be used to control AMR spread (Table 2.3).

2.5 CONCLUSION AND FUTURE PROSPECTS

Antimicrobial resistance is a global threat in view of the present and expected future pollution levels. It is believed that pollutant concentration and AMR in the prevalent microflora are proportional to each other. Along with developing AMR, it is also reported that bacterial colonies also show resistance to heavy metals as the gene for the same lies side by side. Hence it is the possibility that with increasing AMR profile, bacteria may develop resistance to certain other toxic compounds i.e., endocrine disruptors (EDCs), persistent organic pollutants (POPs), pesticides, plasticizers, etc., and will disturb the ecological balance for future generations.

Therefore, the present book chapter is designed to give a glimpse into the vibrant issue of increasing antimicrobial resistance on the global level as well as the wastewater/leachate being their origin of spread. The following conclusions can be drawn from the study:

- AMR spreads sporadically from leachate; 2/3rd of the tested isolates were found to develop multi-drug resistance, which worsens the situation.
- Regular monitoring of antibiotic concentrations in wastewater plants or landfill sites is required to limit their exposure to native microflora.
- Further AMR surveillance at places having a high antibiotic presence can also help in understanding the spectrum.
- The MAR index of isolates is an efficient analytical tool to assess the health risk of indigenous microbes, their dissemination in the environment, and the ultimate threat.

TABLE 2.3
Control Strategies to Minimize the Spread of Antimicrobial Resistance

Control strategies	Contribution
Hygiene and sanitation	Helps restrict the dissemination of antibiotic-resistant microbes
Vaccination	Reduces disease severity in immune-compromised populations, provides immunity against flaking pathogens, and thereby increases the threshold pathogenic load required for infection.
Alternative therapies	Alternative therapies, viz. bacteriophage, probiotics and quorum sensing inhibitors, help in the detection and control of the spread of antibiotic resistance. Doxycycline has exhibited efficacy against MDR *Acinetobacter* in the present study, which can be considered as an alternative treatment apart from carbapenems.
Education	Create awareness among the health care workers and patients about the hazardous health impact of undue usage of antibiotics that leads to an exponential increase of antibiotic resistance in native microflora.
Infection prevention & control	Hospital infection control measures like ducted ventilation, isolation wards, and hygiene maintenance may protect against the spread of nosocomial pathogens and sensitive microbes.

The present study gives a preliminary indication of the spread of AMR from raw as well as recycled leachate, which is supposed to be a very concentrated matrix with a very low organic pollutant load resulting in less microflora. It is expected that in running water bodies, AMR in the native microflora would be significantly higher. Therefore, future studies should focus on studying AMR patterns in different types of water bodies; surface water, sewage, municipal water, etc. to get a clear understanding of the AMR and its interaction with the prevalent pollutant load.

2.6 ACKNOWLEDGMENT

The authors gratefully acknowledge the guidance and support of the Director, CSIR-NEERI, and HOD, CSIR-NEERI, Kolkata Zonal Center. The KRC Knowledge Resource Center of CSIR-National Environmental Engineering Research Institute number for the manuscript is CSIR-NEERI/KRC/2022/OCT/KZC/01.

REFERENCES

Al-Gheethi, A. A., Efaq, A. N., Bala, J. D. et al.2018. Removal of pathogenic bacteria from sewage-treated effluent and biosolids for agricultural purposes. *Appl, Water, Sci.*,8, 74.

Amador, PP., Fernandes, RM., Prudêncio, MC., Barreto, MP., and Duarte, IM. (2015). Antibiotic resistance in wastewater: occurrence and fate of Enterobacteriaceae producers of Class A and Class C b-lactamases. *J. Environ. Sci. Health A* 50, 26–39. doi: 10.1080/10934529.2015.964602

Becker K., Alen S V., Idelevich E A. et al. Plasmid-Encoded Transferable *mecB*-Mediated Methicillin Resistance in *Staphylococcus aureus. Emerg Infect Dis.*, 2018 Feb; 24(2), 242–248. doi: 10.3201/eid2402.171074

Blaak, H. G., Lynch, R., Italiaander, R. A., Hamidjaja, F. M., and de Roda Husman, A. M. 2015. Multidrug-resistant and extended spectrum beta-lactamaseproducing *Escherichia coli* in Dutch surface water and wastewater. *PLoS One*, 10, e0127752. doi: 10.1371/journal.pone.0127752

Blázquez, J., Couce, A., Rodríguez-Beltrán, J., et al. 2012. Antimicrobials as promoters of genetic variation. *Curr. Opin. Microbiol.*, 15,561–569.

Boeckel,Van T. P, Gandra, S., Ashok, A. et al. 2014. Global antibiotic consumption 2000 to 2010: An analysis of national pharmaceutical sales data. *Lancet. Infect. Dis.*, 14, 742–750.

Borquaye, L. S., Ekuadzi, E., Darko, G., Ahor, H.S., et al. 2019. Occurrence of antibiotics and antibiotic-resistant bacteria in landfill sites in Kumasi, Ghana. *JCHEM*, Article ID 6934507, 10 pages.https://doi.org/10.1155/2019/6934507

Brooks, J. P., and McLaughlin, M. R. 2009. Antibiotic resistant bacterial profiles of anaerobic swine lagoon effluent. *J. Environ. Qual*, 38, 2431–2437. doi: 10. 2134/jeq2008.0471

Christopher, A. F., Hora, S., Ali, Z. 2013. Investigation of plasmid profile, antibiotic susceptibility pattern multiple antibiotic resistance index calculation of *Escherichia coli* isolates obtained from different human clinical specimens at tertiary care hospital in Bareilly-India.*ATMPH*, 6,3.

da Silva, M. F., Tiago, I., Veríssimo, A., Boaventura, R. A. R., Nunes, OC., and Manaia, CM. 2006. Antibiotic resistance of *enterococci* and related bacteria in an urban wastewater treatment plant. *FEMS Microbiol. Ecol.*, 55, 322–329. doi: 10.1111/j.1574-6941.2005.00032.x

Garcet, L. A. A., Martinez, J. E. R., Barrer, D. B. V. et al. 2022. Biotechnological potential of micro-organisms from landfill leachate: Isolation, antibiotic resistance and leachate discoloration. *An. Acad. Bras. Cienc.*, 18, 94(3),e20210642.doi: 10.1590/0001-3765202220210642

Garcete, LAA., Martinez, JER., Barrer, DBV. Et al. (2022). Biotechnological potential of micro-organisms from landfill leachate: isolation, antibiotic resistance and leachate discoloration. *An Acad Bras Cienc* 2022 Jul 18;94(3), e20210642. doi: 10.1590/0001-3765202220210642

Kowalska-Krochmal, B.,Dudek-Wicher, R. 2021. The minimum inhibitory concentration of antibiotics: Methods, interpretation, clinical relevance. *Pathogens*, 10(2),165. doi: 10.3390/pathogens10020165

Magiorakos, A. P., Srinivasan, A., Carey, R. B., Carmeli, Y., Falagas, M. E., Giske, C. G., et al. 2012. Multidrug-resistant, extensively drug-resistant and pandrug-resistant bacteria: An international expert proposal for interim standard definitions for acquired resistance. *Clin. Microbiol. Infect.*, 18, 268–281. doi: 10.1111/j.1469-0691.2011.03570.x

Martinez, J.L. 2014. General principles of antibiotic resistance in bacteria. *Drug Discov. Today*, 11,33–39.

Mukherjee, M., Laird, E., Gentry, T.J., Brooks, J. P. and Karthikeyan, R. 2021. Increased antimicrobial and multi-drug resistance downstream of wastewater treatment plants in an urban watershed. *Front. Microbiol.*, 12, 657353. doi: 10.3389/fmicb.2021.657353

Nicolau, D. P., Carmeli, Y., Crank, C. W., Goff, D. A., Graber, C. J., Lima, A. L. L., et al. 2012. Carbapenem stewardship: Does ertapenem affect *Pseudomonas* susceptibility to other carbapenems? A review of the evidence. *Int. J. Antimicrob. Agric.*, 39, 11–15. doi: 10.1016/j.ijantimicag.2011.08.018

Reygaert, W. C. 2018. An overview of the antimicrobial resistance mechanisms of bacteria. *AIMS Microbiol.*, 4(3),482–501. doi: 10.3934/microbiol.3.482

Scaria, J., Anupama, K., Nidheesh, P. 2021. Tetracyclines in the environment: An overview on the occurrence, fate, toxicity, detection, removal methods, and sludge management. *Sci. Total Environ.*, 771, 145291. doi: 10.1016/j.scitotenv

Sekizuka, T., Yatsu, K., Inamine, Y., Segawa, T., Nishio, M., Kishi, N., Kuroda, M. 2018. Complete Genome sequence of a bla_{KPC-2}-positive *Klebsiella pneumoniae* strain isolated from the effluent of an urban sewage treatment plant in Japan. *mSphere*, 19(3), 5.doi:10.1128/mSphere.00314-18

Sivaranjani, V., Umadevi, S., Srirangaraj, S., Kali, A., Seetha, K.S. 2013. Multi drug resistant Acinetobacter species from various clinical samples in a tertiary care hospital from South India. *AMJ*, 6(12), 697–700.

Valitalo, P., Kruglova, A., Mikola, A., Vahala, R. 2017. Toxicological impacts of antibiotics on aquatic micro-organisms: A mini-review. *Int. J. Hyg. Environ. Health.*,220, 558–569. https://doi.org/10.1016/j.ijheh

Wang, YQ,. Tang, W., Qiao, J., Song. L. (2015). Occurrence and prevalence of antibiotic resistance in landfill leachate. *Environ Sci Pollut Res Int.* (16), 12525–33. doi: 10.1007/s11356-015-4514-7

You, X., Wua1,D., Wei, H., Xiea, B., Lud, J. 2018. Fluoroquinolones and β-lactam antibiotics and antibiotic resistance genes in autumn leachates of seven major municipal solid waste landfills in China. *Environ. Int.*, 113, 162–169. doi: 10.1016/j.envint.2018.02.002

Zhao, R., Feng, J., Yin, X. et al. 2018. Antibiotic resistome in landfill leachate from different cities of China deciphered by metagenomic analysis. *Water Res.*, 1(134), 126–139. doi: 10.1016/j.watres.2018.01.063

SUPPLEMENTARY INFORMATION

TABLE S1
Mechanisms of Action Based Classification Different Antibiotics

Sl. No.	Classification	Sub-types	Mechanism of Action against Bacteria	Toxicity	Example Antibiotics
1.	Penicillins	Penicillin	Bactericidal: blocks cross linking via competitive inhibition of the transpeptidase enzyme	Hypersensitivity reaction Hemolytic anemia	Penicillin G Aqueous penicillin G Procaine penicillin G Benzathine penicillin G Penicillin V
		Aminopenicillins		Hypersensitivity reaction Hemolytic anemia	Ampicillin Amoxicillin
		Penicillinase-resistant-penicillins		Hypersensitivity reaction Hemolytic anemia, Interstitial nephritis	Methicillin Nafcillin Oxacillin Cloxacillin Dicloxacillin
		Antipseudomonal penicillins		Hypersensitivity reaction Hemolytic anemia	Carbenicillin Ticarcillin Piperacillin

TABLE S1 (Continued)
Mechanisms of Action Based Classification Different Antibiotics

Sl. No.	Classification	Sub-types	Mechanism of Action against Bacteria	Toxicity	Example Antibiotics
2.	Cephalosporins	1st generation	Bactericidal: inhibits bacterial cell wall synthesis via competitive inhibition of the transpeptidase enzyme	Allergic reaction Coombs-positive anemia (3%)	Cefazolin Cephalexin
		2nd generation		Allergic Reaction ETOH Disulfiram reaction	Cefoxitin Cefaclor Cefuroxime
		3rd generation		Allergic Reaction ETOH Disulfiram reaction	Ceftriaxone Cefotaxime Ceftazidime Cefepime (4th generation)
3.	Other Cell Wall Inhibitors	Vancomycin	Bactericidal: disrupts peptioglycan cross-linkage)	Red man syndrome Nephrotoxicity Ototoxicity	Vancomycin
		Beta-lactamase Inhibitors	Bactericidal: blocking cross-linking	Hypersensitivity Reaction Hemolytic anemia	Clavulanic Acid Sulbactam Tazobactam
		Carbapenems			Imipenem (+ cilastatin) Meropenem Doripenem Ertapenem
		Aztreonam			Aztreonam
		Polymyxins			Polymyxin B Polymyxin E
		Bacitracin			Bacitracin
4.	Protein Synthesis Inhibitor	Aminoglycosides	Bactericidal: irreversible binding to 30S)	Nephrotoxicity Ototoxicity	Gentamicin Neomycin Amikacin Tobramycin Streptomycin
		Tetracyclines	Bacteriostatic: blocks tRNA)	Hepatotoxicity Tooth discoloration Impaired growth Avoid in children < 12 years of age	Tetracycline Doxycycline Minocycline Demeclocycline
		Macrolides	Bacteriostatic: reversibly binds 50S	Coumadin Interaction (cytochrome P450)	Erythromycin Azithromycin Clarithromycin
		Chloramphenicol	Bacteriostatic	Aplastic Anemia Gray Baby Syndrome	Chloramphenicol
		Lincosamide	Bacteriostatic: inhibits peptidyl transferase by interfering with amino acyl-tRNA complex	Pseudomembranous colitis Hypersensitivity Reaction	Clindamycin
		Linezolid	Anti 50S Ribosomal unit		Linezolid
		Streptogramins	Anti 50S Ribosomal unit		Quinupristin Dalfopristin

(*continued*)

TABLE S1 (Continued)
Mechanisms of Action Based Classification Different Antibiotics

Sl. No.	Classification	Sub-types	Mechanism of Action against Bacteria	Toxicity	Example Antibiotics
		1st generation 2nd generation	Fluonoloneroquis- bactericidal: inhibit DNA gyrase enzyme, inhibiting DNA synthesis	Phototoxicity Achilles tendon rupture Impaired fracture healing	Nalidixic acid Ciprofloxacin Norfloxacin Enoxacin Ofloxacin Levofloxacin
		3rd generation 4th generation			Gatifloxacin Moxifloxacin Gemifloxacin
		Metronidazole	Bacteridical: metabolic biproducts disrupt DNA)	Metronidazole	Seizures Crebelar dysfunction ETOH disulfram reaction
5.	RNA synthesis inhibitor	Rifampin	Bactericidal: inhibits RNA transcription by inhibiting RNA polymerase)	Rifampin	Body fluid discoloration Hepatoxicity (with INH)
6.	Mycolic Acids Synthesis Inhibitors	Isoniazid		Isoniazid	
7.	Folic acid Synthesis Inhibitors	Trimethoprim/ Sulfonamides	Bacteriostatic: inhibition with PABA)	Thrombocytopenia Avoid in third trimester of pregnancy	Trimethoprim/ Sulfamethoxazole (SMX) Sulfisoxazole Sulfadiazine
		Pyrimethamine			Pyrimethamine

TABLE S2
Intrinsic Antibiotic Resistance in Bacteria

Organism	Intrinsic Resistance
Bacteroides (anaerobes)	aminoglycosides, many β-lactams, quinolones
All Gram positives	aztreonam
Enterococci	aminoglycosides, cephalosporins, lincosamides
Listeria monocytogenes	cephalosporins
All Gram negatives	glycopeptides, lipopeptides
Escherichia coli	macrolides
Klebsiella spp.	ampicillin
Serratia marcescens	macrolides
Pseudomonas aeruginosa	sulfonamides, ampicillin, 1st and 2nd generation cephalosporins, chloramphenicol, tetracycline
Stenotrophomonas maltophilia	aminoglycosides, β-lactams, carbapenems, quinolones
Acinetobacter spp.	ampicillin, glycopeptides

TABLE S3
Antimicrobial Resistance Mechanisms

Drug	Drug Uptake Limitation	Drug Target Modification	Drug Inactivation	Efflux Pumps
β-Lactams	Decreased numbers of porins, no outer cell wall	Gram pos—alterations in PBPs	Gram pos, Gram neg—β-lactamases	RND
Carbapenems	Changed selectivity of porin			
Cephalosporins	Changed selectivity of porin			
Monobactams				
Penicillins				
Glycopeptides	Thickened cell wall, no outer cell wall	Modified peptidoglycan		
Lipopeptides		Modified net cell surface charge		
Aminoglycosides	Cell wall polarity	Ribosomal mutation, methylation	Aminoglycoside modifying enzymes, acetylation, phosphorylation, adenylation	RND
Tetracyclines	Decreased numbers of porins	Ribosomal protection	Antibiotic modification, oxidation	MFS, RND
Chloramphenicol	Ribosomal methylation	Acetylation of drug	MFS, RND	Chloramphenicol
Lincosamides		Gram pos—ribosomal methylation		ABC, RND
Macrolides		Ribosomal mutation, methylation		ABC, MFS, RND
Oxazolidinones		Ribosomal methylation		RND
Streptogramins				ABC
Fluoroquinolones		Gram neg—DNA gyrase modification Gram pos—topoisomerase IV	Acetylation of drug	MATE, MFS,
Sulfonamides		DHPS reduced binding, overproduction of resistant DHPS		RND
Trimethoprim		DHFR reduced binding, overproduction of DHFR		RND

Source: (Reygaert, 2018).

3 Role of Phytoalexins in Agriculture

Sana Sheikh, Sareen Sheikh, Jyothi Miranda, Vishal Moger, Swaroop Kumar V. S., and Jayalaxmi

3.1 INTRODUCTION

The modern world requires crop species that are stress- and pathogen-free while simultaneously ensuring environmental sustainability. In order to combat pathogenic illnesses of economically significant crop species, natural products are the greatest tools available. The most successful method of sustainable agriculture is to look for solutions in nature. In reaction to radiation exposure, mechanical tissue breakdown, and heavy metal toxicity as abiotic stresses, plants develop low-molecular-weight, secondary metabolites known as phytoalexins. After de novo synthesis of phytoalexin biosynthesizing enzymes, they are synthesized from distant precursors in response to pathogen attack. (Braga, 1991; Echeverri et al., 2012). These secondary metabolites, which are natural substances with a variety of chemical properties, such as alkaloids, coumarins, isoflavonoid (coumestans, isoflavans, isoflavones, pterocarpans, pterocarpenes), lignans, polyacetylenes, and pterocarpons (pisatin, phaseolin, etc.) help the plant withstand biotic and abiotic stress (Grayer and Kokubun, 2001). They are the first choices for agriculturalists because of their natural tendency to degrade.

3.1.1 Plant Protection with Phytoalexins

When stimulated by certain phytopathogenic microbes or by mechanical or chemical harm, plants create significant amounts of poisonous antimicrobial molecules referred to as phytoalexins (Singh et al., 2017). Antimicrobial compounds that are synthesized in plants can be broadly classified into two categories, phytoantipicins and phytoalexins. Phytoanticipins are "low molecular weight," antimicrobial compounds that are present in plants before invasion by microbes, or are produced after infection solely, from pre-existing precursors, while a broad class of biochemical defence compounds, termed phytoalexins, are loosely defined as any "low molecular weight, antimicrobial compounds that are both synthesized and accumulated in plants after exposure to microbes or abiotic agents." They may be biosynthetically derived from one or several primary biosynthetic pathways (Singh and Chandrawat, 2017), whereas leaves, stems, and flowers are sites for phytoalexin biosynthesis.

From roughly 30 plant families, more than 350 phytoalexins have been chemically characterized. A total of 130 Leguminosae species have been characterized. The majority of phytoalexins made by the Leguminosae are classified into six isoflavonoid classes: coumestans, isoflavones, isoflavanones, pterocarpans, and pterocarpenes. Pisatin, phaseollin, glyceollin, medicarpin, and maackiain are a few Pterocarpan phytoalexins. The garden pea, *Pisum sativum*, was the source of the first phytoalexin to be isolated and characterized. In addition to these substances, some legumes can produce

DOI: 10.1201/9781003268895-3

non-isoflavonoid phytoalexins such as stilbenes and furanoacetylenes (Alghisi and Favaron, 1995; Ahuja et al., 2012).

Elicitors are the chemicals that alert plants to start the process of phytoalexin production. Plants and potential pathogens may interact with elicitors of biotic origin, whereas abiotic elicitors are not typically involved in host-pathogen interactions (Singh and Chandrawat, 2017). The development of pathogenic resistance is caused by the accumulation of phytoalexins in the cells surrounding the infected areas, which is induced by contact with a pathogen in the form of a bacteria or fungus through an infection. This further activates biosynthetic pathways, which in turn leads to its transcription. There are a few other abiotic variables that can cause similar effects (Taiz and Zeiger, 2002).

3.2 PRODUCTION OF PHYTOALEXIN

The first instance of phytoalexin production to be documented was "late blight of potato," which was brought on by *Phytophthora infestans*. When infected with another strain of the same Phytophthora genus, this pathogenic fungus triggered the hypersensitive reactions that resulted in the creation of a "plant secondary metabolite" that prevented continued infection in the same plant. These substances, which Muller and his coworkers dubbed "phytoalexins," have shielded the plant from secondary infection (Deverall, 1982).

As the infection progresses, the pathogen creates proteins and carbohydrates that may cause plants to respond by producing and accumulating phytoalexins. Elicitor and elicitin are two different types of inducers. The oligosaccharides from fungal cells that act as elicitors are frequently used. The glycoproteins known as elicitins are typically secreted by the fungal cells (Cordelier et al., 2003).

Plant infections are also responsible for the biological degradation of phytoalexins. For example, resveratrol, which protects grape plants from pathogens by accumulating at high concentrations, is degraded by *Botrytis cinerea* (Jeandet et al., 2021). For a variety of host plant families, including Cruciferae, Solanaceae, Graminae, and Leguminosae, phytopathogenic fungi have been reported to metabolize and detoxify a number of phytoalexins (Sobolev et al., 2022) (Table 3.1).

3.2.1 Compounds Involved in Plant Defence

Some of the substances involved in plant defence include saponins, cyanogenic glycosides, and gucosinolates.

3.2.1.1 Saponins

According to the structure of the aglycone, which may be a triterpenoid, a steroid, or a steroidal glycoalkaloid, saponins are glycosylated chemicals that are commonly found in the plant world and can be split into three broad classes. Triterpenoid saponins are mostly found in dicotyledonous plants, while they can also be found in some monocots. Steroid saponins, on the other hand, are

TABLE 3.1
Sources of Phytoalexins

Phytoalexins	Sources
Rishitin	Potato, Tomato
Capsidiol	Pepper
Gossypol	Cotton
Glyceollins	Soyabean
Pisatin	Garden Pea

mostly found in monocots, such as Liliaceae, Dioscoraceae, and Agavaceae, and in some dicots, like foxglove, which has the saponin digitonin (Hostettmann and Marston, 1995). Oats (genus *Avena)* differ from other plants in that they contain triterpenoid and steroid saponins. The family Solanaceae, which includes plants like potatoes and tomatoes, as well as the Liliaceae, are the main sources of steroidal glycoalkaloids (Hostettmann and Marston, 1995). In terms of their possible function in defending plants against phytopathogenic fungi, the saponins generated by oats and tomatoes have been the subjects of the most extensive research (Osbourn, 1996).

In general, the following saponins have a role in plant defence:

3.2.1.1.1 Avenacins and Avenacosides

The triterpenoid avenacins and the steroidal avenacosides are two distinct groups of saponins that are present in oats. Both avenacins and avenacosides are forms of saponin that are distributed differently throughout the plant; avenacins are found in the root (Goodwin and Pollock, 1954), whereas avenacosides are found in the leaves and shoots (Tschesche et al., 1969). The role of the triterpenoid avenacin saponin in determining oats' resistance to the "take-all" disease-causing root-infecting fungus *Gaeumannomyces graminis* var *tritici* has attracted a lot of attention. While *G.g. tritici* severely reduces wheat and barley yields, it is unable to infect oats. Additionally, unlike the *G. graminis* variation that attacks oats, *G. graminis* var *avenae*, *G. g. tritici* is particularly susceptible to avenacins. As a result, the presence of these saponins in oat roots has been suggested as the cause of oats' resistance to *G. g. tritici* (Turner, 1953).

3.2.1.1.2 Tomato Saponin, a-Tomatine

A-tomatine, a steroidal glycoalkaloid, is the main saponin found in tomatoes. Similar to avenacins, a-tomatine is monodesmosidic and found in healthy plants in its biologically active form. The tomato leaves, flowers, and green fruits have exceptionally high concentrations of this saponin (Roddick, 1974).

3.2.1.1.3 Cyanogenic Glycosides

In reaction to tissue damage, more than 2000 plant species are known to be cyanogenic and capable of producing hydrocyanic acid (HCN) (Davis, 1991). HCN has typically been linked to defence against herbivores and infections because of its toxicity to most species, particularly as a respiratory toxin (Hughes, 1991). Plant hydrolytic enzymes degrade precursors, typically cyanogenic glycosides, and as a result, HCN is released (Poulton, 1988). Amygdalin, the most well-known of them, is in charge of the toxicity of bitter almonds and other rosaceous nuts. Cyanoglycosides are also known to be present in a variety of grasses, including cereals; dhurrin, a substance produced by sorghum species, is one particularly well-known example.

3.2.1.1.4 Glucosinolates

The family Cruciferae, which includes the cruciferous weed Arabidopsis and the crucially important species Brassica, contains glucosides called glucosinolates, which contain sulphur. The primary relevance of these substances to Brassica's interactions with vertebrate and invertebrate pests is in their capacities as attractants or feeding inhibitors (Mithen, 1992; Giamoustaris and Mithen, 1995). Tissue-specific glucosinolate dispersion occurs within individual plants. For instance, indolyl and phenylethyl glucosinolates are the main glucosinolates in the roots and stems of oilseed rape, whereas aliphatic glucosinolates prevail in the leaves of the plant (Mithen, 1992).

3.3 FUNCTIONAL ROLE OF PHYTOALEXINS

Phytoalexins have potent antimicrobial and medicinal activities. As such, scientists have attempted to increase their accessibility by inventing diverse in vitro and in vivo approaches. Among these approaches, bioengineering plant transcription factors that directly regulate phytoalexin biosynthesis genes may be the most promising. Recent research has identified conserved transcription factors that directly regulate distinct phytoalexin biosynthesis pathways in different plant species. The interesting results provide new insight into how the defence signaling pathways in plants result in lineage-specific biochemical defences. These recent findings also suggest that a common transcription factor network could be engineered to enhance the biosynthesis of different phytoalexins in plants (Ahmed and Kovinich, 2020).

Numerous studies on their use in agricultural fields have been done due to their beneficial effects on plant health. Many of these chemicals have been found to be effective against their pathogenic hosts in agricultural crops even though many of them are less effective than chemical fungicides at combating fungal pathogens, they may nevertheless be preferable to chemical fungicides (Jeandet et al., 2014).

Investigations on some selected number of species and genera are made from plant families including both monocotyledons (Amaryllidaceae, Orchidaceae, Poaceae) and dicotyledon plants (Apiaceae, Asteraceae, Convolvulaceae, Chenopodiaceae, Euphorbiaceae, Linaceae, Moraceae, Piperaceae, Rosaceae, Rutaceae, Leguminosae and Solanaceae) for their phytoalexin production (Ingham, 1982; Coxon, 1982). Few plant species of economically important families such as maize oats, sorghum and rice (Poaceae), cotton (Malvaceae), and Vitaceae species also produce phytoalexins as a mode of inducible defence (Jeandet et al., 2010; Jeandet et al., 2014; Schmelz et al., 2014). "Camalexin" produced by members of the family Brassicaceae has been one of the important phytoalexins on which numerous studies focusing on its biosynthetic pathway and regulatory factors involved in its production have been carried out specifically on model plant *Arabidopsis thaliana* (Ren et al., 2008).

Phytoalexins accumulate at infection sites and they inhibit the growth of fungi and bacteria in vitro therefore, it is logical to consider them as possible plant-defence compounds against diseases caused by fungi and bacteria. Phytoalexins are considerably less toxic than chemical fungicides. Phytoalexin fungitoxicity is clearly evidenced by the inhibition of germ-tube elongation, radial mycelial growth, and mycelia dry weight increase, as best illustrated by the action of resveratrol on *Botrytis* cinerea, the causal agent for grey mould in grapevine. Phytoalexins may also exert some effects on the cytological, morphological, and physiological characteristics of fungal cells (Alghisi and Favaron, 1995).

Knowledge of control mechanisms involved in the production, accumulation, and regulation of phytoalexins in different systems could serve as a basis for genetic manipulation of these compounds in genetically engineered plants for enhanced disease resistance (Jeandet et al., 2012). Phytoalexins are chemically diverse compounds; for example, in the family Fabaceae six isoflavonoid classes have been reported: isoflavanones, isoflavones, pterocarpenes, pterocarpans, coumestans, and isoflavans. Well-known pterocarpans are phaseolin, glyceollin, pisatin, maackiain, and medicarpin. Legumes also produce nonisoflavonoid phytoalexins such as stilbenes and furanoacetylenes . In the family Solanaceae five different classes of phytoalexins have been reported: steroid glycoalkaloids, sesquiterpenoids and nonsesquiterpenoids, phenylpropanoid-related compounds, polyacetylene derivatives, and coumarins. Two new phytoalexin classes have also been discovered in members of the Brassicaceae and Poaceae family. In rice, maize, and sorghum zealexins, kauralexins, momilactones, phytocassanes, and oryzalexins are reported (Pedras et al., 2000).

Scientists have proven that not only infection acts as the stimulus for phytoalexin synthesis but also the hormone levels; phosphorylation cascades play a major role in this. Cytokinin overexpression in *Nicotiana tobaccum* is directly associated with its resistance against *P. syringe* by higher concentrations of capsidiol and scopoletin (Grosskinsky et al., 2011). Kauralexin biosynthesis seems to be regulated by phytohormones in maize. The application of a combination of jasmonic acid (JA) and ethylene to maize plants was sufficient to induce kauralexin generation. The gene AN2 was highly expressed at 24 hours after an attack by the insect *Ostrinia nubilalis*, which was accompanied by an increase in the concentrations of JA and ethylene together with the expression of allene oxide synthase and 1-aminocyclopropane-1-carboxylic acid oxidase genes, key enzymes involved in biosynthesis of these hormones (Poloni and Schirawski, 2014).

3.3.1 Functional Role of Phytoalexins against Pathogens in Agriculture

The chemical nature of phytoalexins was mainly shown to be different classes of isoflavanoids in the family Leguminosae and in the family Solanaceae it was mainly terpenoids. These substances had a variety of impacts on the several fungal infections implicated, including decreased glucose uptake, membrane disruption, and asymmetric development of germ tubes (Singh and Chandrawat, 2017). Studies of plant disease resistance mechanisms have recently tended to concentrate on the production of phytoalexins and other active responses brought on by pathogen infection (Hammond-Kosack and Jones, 1996).

The effectiveness of phytoalexins in sorghum and maize in preventing bacterial and fungal diseases was examined (Poloni and Schirawski, 2014). When *colletotrichum sublineolum* was used to stimulate the production of two different forms of 3-deoxyanthocyanidin phytoalexins, it exhibited a reduction in the proliferation and dissemination of fungal hyphae. Additionally, it was shown that the biosynthesis of these substances affected the formation of flavones and anthocyanins. Reports also showed that apigeninidin, one of the 3-deoxyanthocyanins, inhibited growth of many Gram-positive bacteria as well as the Gram-negative bacteria. Additionally, studies revealed that the 3-deoxyanthocyanin apigeninidins hindered the growth of numerous Gram-positive and Gram-negative bacteria and revealed the potential for creating disease-resistant plants that produce a variety of phytoalexins by biotechnology or conventional breeding techniques, which would help prevent losses in the field.

Morandi (1996) studied the papers showing the effect of endomycorrhizal infections in increasing phytoalexins, phenolics, and associated molecules in roots. The thorough study however showed that a group of phytoalexins, the isoflavanoids showed an increase in in-vitro mycorrhizal infection of roots. Also, several papers showed flavonols were more efficient in stimulating hyphal elongation of mycorrhizal than flavones and flavanones. Reviews of in-vitro studies revealed that mycorrhizal soybeans were more tolerant of nematodes than non-mycorrhizal ones due to the accumulation of the nematotoxic phytoalexin glyceollin. As a result, this could be employed in agriculture for efficient biocontrol, although further research is needed to fully understand these relationships.

Langcake and Pryce (1977) isolated another class of phytoalexins from *Vitis vinifera.* These classes of compounds were known as viniferins through the infection of leaves by the fungus *Botrytis cinerea.* The chemical characterization of this compound was done and it was found to be closely related to trans-resveratrol. The antifungal activities of resveratrol, when compared to viniferins were lower and were proven through spore germination tests.

Jeandet et al. (2014) showed the potentiality of phytoalexins in strengthening food and agriculture using molecular engineering. The introduction of stilbene synthase gene to produce revestraol was shown to increase resistance to several diseases in a wide variety of host plants. In some plants like tobacco, it was observed that the strategy of engineering phytoalexin synthesis enzymes in plants may produce weaker resistance due to phytoalexins and be less efficient than chemical fungicides. In food industries, research is conducted mainly to reduce the phytoalexins in certain plants due to their toxicity to humans at their present levels. However, such modifications could weaken the plant's inherent capacity to survive adverse circumstances.

The reviews on the biological and chemical features of phytoalexins were compiled (Grayer and Kokubun, 2001). The search for phytoalexins has taken place in different families like Leguminosae, Vitaceae, Solanaceae, and Rosaceae. The majority of the substances in this group are either terpenoids or isoflavanoids. Biphenyls and dibenzofurans were found to be the chemicals in the Rosaceae family that triggered antifungal activities. The development of rice varieties resistant to rice blast disease that are based on phytoalexins was also revealed. The authors also explained the nuances of extracting phytoalexins in terms of its high expense and also the marked toxicity of these phytoalexins against other organisms including humans. The approach of immunizing the plants by inoculating them with a weakened non-pathogenic microorganism or an incompatible strain of a pathogen to elicit resistance against the actual pathogen by accumulation of the phytoalexins.

Ahuja et al. (2012) evaluated the presence of phytoalexins in many different families of crop plants like Brassicaceae, Fabaceae, Solanaceae, Vitaceae, and Poaceae. The alkaloids and indole compounds like camalexin were shown to be phytoalexins in Cruciferae. Many of these compounds were shown to be resistant to fungi and also a few bacteria. The authors also assessed the issue of lack of understanding of some aspects of their biosynthesis pathway that might hinder the approach towards developing such resistant strains of these crops.

Schmelz et al. (2014) investigated the impact of terpenoid phytoalexins in monocots, particularly rice (*Oryza sativa*), maize *(Zea maize)*, and wheat (*Triticum* spp). In rice and maize, diterpenoid phytoalexin acts as an antibacterial phytoalexin. They focused on the discovery, elicitation of biosynthesis, and functional characterization of a monocot terpenoid phytoalexin found in rice, maize, and wheat.

Sobolev et al. (2011) worked on the biological activity of peanut phytoalexin. When a microbial pathogen infects the peanut plant (*Arachis hypogaea* L.), the plant can produce chemicals called antifungal phytoalexins that are derived from stilbene. Several researchers have also identified the resveratrol and pterostilbene stilbenoids from peanuts as having possible health advantages. In a variety of biological experiments, this study looked into how some of these chemicals behaved. Since peanut stilbenoids seem to be involved in plant defence systems, their impact on commercially significant plant pathogenic fungi belonging to the genera *Colletotrichum*, *Botrytis*, *Fusarium*, and *Phomopsis* was examined. These studies' findings demonstrate that peanut stilbenoids, along with related natural and synthetic stilbene derivatives, have a variety of biological functions.

"Fungal Elicitors of Phytoalexins and their Potential use in Agriculture" was the focus of Paxton et al. (1998). Plant phytoalexin elicitors from fungi have the potential to develop into a new class of herbicides and plant protectants. The Oomycete *Phytophthora megasperma* f. sp. glycinea cell walls contain the elicitor that has currently been most thoroughly characterized. A heptaglucan with specified structural criteria is the smallest active component of this elicitor and may also be the most helpful one. They discovered that although other fungal polysaccharides may not have as well-characterized structures, they also cause plants to produce phytoalexin. Phytoalexin elicitation has also been linked to fungus pectinases. These enzymes appear to cause plants to produce phytoalexins by releasing pectic particles from their cell walls. They came to the conclusion that using elicitors in agriculture has tremendous potential. Because phytoalexins play a significant part in plant disease and pest resistance, their controlled elicitation may be utilized to promote natural disease and pest resistance without the use of substances that harm the environment.

Several fungal infections, including *Penicilium digitatum*, *Magnaporthae oryzae*, *Phytophthora* species, *Rhizoctonia solani*, *Colletotrichum graminicola*, *Fusarium* species, *Puccina coronta*, and *Botrytis cinerea*, have been demonstrated to be resistant to a variety of phytoalexins. Additionally, at a concentration of 360 g/L, the phytoalexin rishitin reduced the viability of *Erwinia atroseptica* bacterial cells by almost 100% (Ahmed and Kovinich, 2020).

3.3.1.1 Phytoalexin in Sorghum

Apigeninidin and luteolinidin are two different 3-deoxyanthocyanidin phytoalexins that are produced by sorghum. Various compounds, such as 5-methoxy-luteolindin, the caffeic acid ester of Arabinosyl-5-O-apigeninidin, and 7-methoxyapigeninidin, are also present. It is known that the cytoplasm of epidermal sorghum cells infected with *Colletotrichum sublineolum* produces a mixture of these distinctive reddish and orange coloured chemicals, which collect in at first colourless inclusion bodies. These inclusion bodies travel to the infection zone, where they first build up and turn pigmented, lose their spherical shape, and then release their red contents at the infection site. Resistance cultivars accumulate the 3-deoxyanthocyanidins significantly more quickly than susceptibility cultivars do when exposed to pathogens. To stop the growth and spread of fungal hyphae, early phytoalexin accumulation is crucial (Poloni, 2014).

3.3.1.2 Phytoalexin in Maize

Terpenoids, such as zealexins and kauralexins, and benzoxazinoids, such as DIMBOA (4-dihydroxy-7-methoxy-1,4-benzoxazin-3-one-glucoside) and HDMBOA (2-hydroxy-4,7-dimethoxy-1,4-benzoxazin-3-one-glucoside), are the phytoalexins found in maize. Recent research has revealed zealexins to be a class of -macrocarpene-related acidic sesquiterpenoids. Zealexin A1, A2, A3, B1, and C3 are among at least five distinct substances that belong to the group that build up in *Fusarium graminearum*, the pathogen of maize, to extremely high levels of roughly 800 μg/g. Maize that had been exposed to other fungal pathogens like *Aspergillus flavus* and *Rhizopus microsporus*, also contained high quantities of zealexins. Zealexin A1 and A3 (but not A2) suppressed the growth of *Aspergillus flavus* and *Fusarium graminearum*, while A1 was also efficient against *Rhizopus microsporus*, according *to in vitro* toxicity studies (Poloni, 2014).

3.4 BIOSYNTHETIC PATHWAYS AND REGULATORY MECHANISM

Numerous compounds come from one or more distinct biosynthetic pathways. However, within a specific family, phytoalexins always have the same chemical makeup. As a result, they have served as chemotaxonomic markers for research on evolutionary relationships. Researchers have found phytoalexins from the Leguminosae, Solanaceae, Amaryllidaceae, Euphorbiaceae, Orchidaceae, Chenopodiaceae, Compositae, Convolvulaceae, Ginkgoaceae, Poaceae, Linaceae, Moraceae, Orchidaceae, Piperaceae, Rosaceae, Rutaceae (Ahmed and Kovinich, 2020).

Different biochemical processes are employed to produce different kinds of phytoalexins. The methylerythritol phosphate (MEP) and geranylgeranyl diphosphate (GGDP) route, the phenylpropanoic-polymalonic acid method, and the indole phytoalexin (IP) pathway are the three most prevalent pathways (Jeandet et al., 2014). Numerous investigations have uncovered the intricate biochemical machinery behind their de novo synthesis, regulation, and mode of action (Jeandet et al., 2013 Ahuja et al., 2012). There are reports on genetic engineering-mediated manipulation of phytoalexin production and increased disease resistance in the case of plants (Delaunois et al., 2009; Jeandet et al., 2013).

After recognizing the particle either from damaged cells or from pathogens, the plant cell launches two forms of induced resistance. First, plants use reactive oxygen species like hydrogen peroxide and superoxide to kill invading cells as a short-term response. When interacting with a pathogen, cells near the infection site are signalled to undergo apoptosis (programmed cell death), which prevents the disease from spreading to other plant tissues in the case of a short-term response or hypersensitive response. The alternative, known as long-term resistance/systemic acquired resistance (SAR), entails damaged plant tissues communicating with the rest of the plant utilizing plant hormones like ethylene, jasmonic acid, salicylic acid, or abscisic acid. Global changes inside the plant take place after receiving the hormonal signal from the injured tissues, such as stimulating specific genes to translate proteins essential in providing defence against encroaching pathogens and also synthesizing enzymes necessary for the manufacture of phytoalexins (Tiku, 2020).

Flavonoid 30 -hydroxylase encoded by SbF30 H2 gene in sorghum expresses the production of phytoalexin 3-deoxyanthocyanidin (Shih et al., 2006). To keep up with the pathogenic evolution, genes that produce phytoalexins have also been genetically altered. The detailed biochemical analysis of the phytoalexins and study of their regulatory mechanisms are opening up new horizons for the universal use of phytoalexin-inducing elicitors as plant defence enhancers.

3.5 CONCLUSION

One element of the intricate systems for plant disease resistance is phytoalexins. Phytoalexins play an important role in plant defence mechanisms, combating diseases and pests. A deeper understanding of phytoalexins would contribute to better development of applications towards sustaining agriculture. Future studies on these intriguing phytoalexins will contribute to the preservation of global nutrition and the advancement of the global economy in a variety of ways.

REFERENCES

Ahmed, S., Kovinich, N. (2020). Regulation of phytoalexin biosynthesis for agriculture and human health. *Phytochemistry Reviews*. doi:10.1007/s11101-020-09691-8

Ahuja, I, Kissen, R., Bones, A. M. (2012). Phytoalexins in defense against pathogens. *Trends Plant Sci.*, 17: 73–90.

Alghisi, P., Favaron, F. (1995). Pectin-degrading enzymes and plant parasite interactions. *Euro J Plant Pathol.*, 101: 365–375.

Braga, M. R. (1991). Phytoalexins induction in Rubiaceae. *J Chem Ecol.*, 17: 1079–1090.

Cordelier, S., Ruffray, P., Fritg, B., Kauffmann, S. (2003). Biological and molecular comparison between localized and systemic acquired resistance induced in tobacco by *Phytophthora megasperma* glycoprotein elicitin. *Plant Mol Biol.*, 51: 109–118.

Coxon, D. T. (1982). Phytoalexins from Other Plant Families. In: Bailey J.A, Mansfield J.W. (eds) *Phytoalexins*. Glasgow/London: Blackie, pp. 106–132.

Davis, R. H. (1991). Glucosinolates. In: DMello, J.P., Duffus, C.M., Duffus, J.H. (eds) *Toxic Substances in Crop Plants*, Cambridge, UK: Royal Society of Chemistry, pp. 202–225.

Delaunois, B., Cordelier, S., Conreux, A., Clément, C., Jeandet, P. (2009). Molecular engineering of resveratrol in plants. *Plant Biotechnol J.*, 7: 2–12.

Deverall, B. J. (1982). Introduction. In: Bailey J.A, Mansfield J.W. (eds) *Phytoalexins*. Glasgow/London: Blackie, pp. 19–20.

Echeverri, F., Torres, F., Quiñones, W., Escobar, G., Archbold, R. (2012). Phenylphenalenone phytoalexins, will they be a new type of fungicide? *Phytochem Rev.*, 11: 1–12.

Giamoustaris, A., Mithen, R. (1995). The effect of modifying the glucosinolate content of leaves of oilseed rape (Brassica napus sp. oleifera) on its interaction with specialist and generalist pests. *Ann. Appl. Biol.*, 126, 347–363.

Goodwin, R. H., Pollock, B. M. (1954). Studies on roots. I. Properties and distribution of fluorescent constituents in Avena roots. *Am. J. Bot.* 4, 516–520.

Grayer, R. J, Kokubun, T. (2001) Plant-fungal interactions: The search for phytoalexins and other antifungal compounds from higher plants. *Phytochem.*, 56: 253–263.

Grosskinsky, D. K, Naseem, M., Abdelmoshem, U. A., Plickert, N., Engelke, T., Griebel, T., Zeier, J., Novak, O., Strand, M., Pfeifhofer, H. (2011). Cytokinins mediate resistance against *Pseudomonas syringae* in tobacco through increased antimicrobial phytoalexin synthesis independent of salicylic acid signaling. *Plant Physiol.*, 2011(157): 815–830.

Hammond-Kosack, K. E., Jones, J. D. G. (1996). Resistance gene-dependent plant defense responses. *Plant Cell*, 8, 1773–1791.

Hostettmann, K. A., Manton, A. (1995). *Saponins. Chemistry and Pharmacology of Natural Products*. Cambridge, UK: Cambridge University Press.

Hughes, M. A. (1991). The cyanogenic polymorphism in Tfolium repens L. (white clover). *Heredity*, 66, 105–115.

Ingham, J. L. (1982). Phytoalexins from the Leguminosae. In: Bailey J.A., Mansfield J.W. (eds) *Phytoalexins*. Glasgow/London: Blackie, pp. 21–80.

Jeandet, P., Delaunois, B., Conreux, A., Donnez, D., Nuzzo, V., Cordelier, S., Clément, C., Courot, E. (2010) Biosynthesis, metabolism, molecular engineering and biological functions of stilbene phytoalexins in plants. *Biofactors*, 36: 331–341.

Jeandet, P., Delaunois, B., Aziz, A., Donnez, D., Vasserot, Y., Cordelier, S., Courot, E. (2012). Metabolic engineering of yeast and plants for the production of the biologically active hydroxystilbene, resveratrol. *J Biomed Biotechnol.*, 2012: 579089.

Jeandet, P., Clément, C., Courot, E., Cordelier, S. (2013). Modulation of phytoalexin biosynthesis in engineered plants for disease resistance. *Int J Mol Sci.*, 14: 14136–14170.

Jeandet, P., Hébrard, C., Deville, M-A., Cordelier, S., Dorey, S., Aziz, A., Crouzet, J. (2014). Deciphering the role of phytoalexins in plant-microorganism interactions and human health. *Molecules*, 19: 18033–18056.

Jeandet, P., Vannozzi, A., Sobarzo-Sanchez, E., Uddin, M. S., Bru, R., Martinez-Marquez, A., Clément, C., Cordelier, S., Manayi, A., Nabavi, S., F., Rasekhian, M., Batiha, G. E. S., Khan, H., Morkunas, I., Belwal, T., Jiang, J., Koffas, M., Nabavi, S. M. (2021). Phytostilbenes as agrochemicals: Biosynthesis, bioactivity, metabolic engineering and biotechnology. *Nat. Prod. Rep.*, 28: 1282–1329.

Langcake, P., Pryce, R. (1977). The production of resveratrol and the viniferins by grapevines in response to ultraviolet irradiation, *Phytochem.*, 16(8): 1193–1196.

Mithen, R. (1992). Leaf glucosinolate profiles and their relationship to pest and disease resistance in oilseed rape. *Euphytica*, 63: 71–83.

Morandi, D. (1996). Occurrence of phytoalexins and phenolic compounds in endomycorrhizal interactions, and their potential role in biological control. *Plant Soil*, 185: 241–251. https://doi.org/10.1007/BF02257529

Osbourn, A. E. (1996). Preformed antimicrobial compounds and plant defense against fungal attack. *Plant Cell*, 8(10): 1821–1831. https://doi.org/10.1105/tpc.8.10.1821

Paxton, J. D. (1988). "Fungal Elicitors of Phytoalexins and Their potential use in agriculture," in Acs symposium series. *J. Am. Chem. Soc.*, 109–119. Available at: https://doi.org/10.1021/bk-1988-0380.ch008

Pedras, M. S. C., Okanga, F. I., Zaharia, I. L., Khan, A. G (2000). Phytoalexins from crucifers: Synthesis, biosynthesis and biotransformation. *Phytochem.*, 53: 161–176.

Poloni, A., Schirawski, J. (2014). Red card for pathogens: Phytoalexins in Sorghum and Maize. *Molecules*, 19: 9114–9133.

Poulton, J. E. (1988). Localisation and Catabolism of Cyanogenic Glycosides. In: Evered, D., Harnett, S. (eds) *Cyanide Compounds in Biology. Proceedings of the ClBA Foundation Symposium* 140, Chichester, UK: John Wiley, pp. 67–91.

Ren, D., Liu, Y., Yang, K. Y., Han, L., Mao, G., Glazebrook, J., Zhang, S. (2008). A fungal-responsiveMAPK cascade regulates phytoalexin biosynthesis in Arabidopsis. *Proc Natl Acad Sci U S A.*, 105: 5638–5643.

Roddick, J. G. (1974). The steroidal glycoalkaloid a-tomatine. *Phytochem.*, 13, 9–25.

Schmelz, E. A, Huffaker, A., Sims, J. W., Christensen, S. A, Lu, X., Okada, K., Peters, R. J. (2014). Biosynthesis, elicitation and roles of monocot terpenoid phytoalexins. *Plant J.*, 79: 659–678. https://doi.org/10.1111/tpj.12436

Shih, C-H., Chu, I., Yip, W., Lo, C. (2006). Differential expression of two flavonoid 30 -hydroxylase cDNAs involved in biosynthesis of anthocyanin pigments and 3-deoxyanthocyanidin phytoalexins in sorghum. *Plant Cell Physiol.*, 47: 1412–141.

Singh, R., Chandrawat, K. S. (2017). "Role of Phytoalexins in plant disease resistance," *Int J Curr Microbiol Appl Sci*, 6(1): 125–129. https://doi.org/10.20546/ijcmas.2017.601.016

Sobolev, V. S. et al. (2011). "Biological activity of peanut (Arachis hypogaea) Phytoalexins and selected natural and synthetic stilbenoids," *J. Agric. Food Chem.*, 59(5), 1673–1682. Available at: https://doi.org/10.1021/jf104742n

Sobolev, V. S., Walk, T. E., Arias, R. S., Massa, A. N., Orner, V. A., Lamb, M. C. (2022). Transformation of major peanut (Arachis hypogaea) Stilbenoid Phytoalexins caused by selected microorganisms. *J. Agric. Food Chem.*, 70(4), 1101–1110.

Taiz, L., Zeiger, E. (2002). *Plant Physiology*. 3rd. England: Sinauer Associates.

Tiku, A. R. (2020). Antimicrobial Compounds (Phytoanticipins and Phytoalexins) and Their Role in Plant Defense. In: Mérillon, J.M., Ramawat, K. (eds) *Co-Evolution of Secondary Metabolites. Reference Series in Phytochemistry*. Cham: Springer. https://doi.org/10.1007/978-3-319-96397-6_63

Tschesche, R., Tauscher, M., Fehlhaber, H. W., Wulff, G. (1969). Avenacosid A, ein bisdesmosidisches Steroidsaponin aus Avena sativa. *Chem. Eer.* 102, 2072–2082.

Turner, E. M. (1953). The nature of resistance of oats to the take-all fungus. *J. Exp. Bot.* 4, 264–271.

4 Livestock Manure Application Causes the Spread of Antibiotic-resistant Genes in Agricultural Lands

*Dileep Kumar, Anam Aslam, Ranjana Gautam, Ankita Gupta, Anurag Yadav, and Kusum Yadav**

*Corresponding author (yadav_k@lkouniv.ac.in, anukusum@gmail.com)

4.1 INTRODUCTION

In response to the rising demand for animal meals, intensive and large-scale livestock agriculture produces massive amounts of animal waste manure (Wu et al., 2023). This livestock waste manure is frequently used as a soil fertilizer in agricultural practices worldwide (Jauregi et al., 2021). However, recent findings have revealed a concerning correlation between using livestock manure as fertilizer and the spread of antibiotic-resistant genes (ARGs) in agricultural lands (Jauregi et al., 2021). The manure employed to control and treat animal infectious diseases contains enormous concentrations of antibiotics from livestock farms (Wu et al., 2023). This manure can carry antibiotics, antibiotic-resistance genes, drug residues, and antibiotic-resistant bacteria to the agricultural land, leading to potential risks of antimicrobial resistance (AMR) dissemination (Blau et al., 2019). Horizontal gene transfer, the method by which genetic material is transferred between organisms without reproductive technology, poses a significant concern in expanding antibiotic-resistant genes in agricultural areas (Ahmad et al., 2021). This process can create bacteria resistant to multiple antibiotics, making it more challenging to treat diseases caused by these microorganisms (Hernando-Amado et al., 2019). Livestock manure is commonly used as a fertilizer in intensive agriculture systems, resulting in high levels of antibiotics and antibiotic-resistant bacteria (Manyi-Loh et al., 2018). Reducing antibiotic use in livestock farming is crucial to combatting the spread of antibiotic-resistant genes in agricultural lands (Manyi-Loh et al., 2023). Additionally, adopting non-livestock manure-based fertilizer application methods and employing superior manure management strategies like composting can help reduce the spread of antibiotic-resistant genes (Keenum et al., 2021). The alarming increase in antibiotic-resistant diseases worldwide underscores the complex nature of the issue (O'Neil, 2014; Xiao et al., 2016; Ranjalkar et al., 2019; Van Boeckel et al., 2019). Most of the world's antibiotic use comes from animal husbandry, emphasizing the need to carefully manage antibiotic use in this sector (Van Boeckel et al., 2017). Antibiotic-resistant genes are naturally present in the soil, and the use of animal medications contributes to the growth of antibiotic-resistant bacteria in manures and the environment (Sharma et al., 2016). Horizontal gene transfer, facilitated by mobile genetic elements in antibiotic-resistant bacteria, plays a crucial role in the rapid spread of antibiotic resistance (Vinayamohan et al., 2022; Che et al., 2019; Karkman et al., 2018). Various factors influence the fate of antibiotics in agricultural soil, including sorption to organic particles, degradation/transformation, and physicochemical properties of antibiotics and

DOI: 10.1201/9781003268895-4

soil (Kummerer, 2004). Soil organic matter content and pH are critical in antibiotic sorption, particularly for sulfonamides (Wegst-Uhrich et al., 2014). Hydrophobic interactions and the presence of organic matter derived from manure can influence the sorption of sulfonamides (Figueroa Diva et al., 2010; Zhang et al., 2011). The widespread and increasing use of antibiotics in food animal production poses environmental dangers and raises concerns about AMR transmission to human and animal sectors (Pérez-Valera et al., 2019). Antibiotic accumulation in crops for human and animal consumption is an emerging issue under investigation (Safdar, 2020).

4.2 DISSEMINATION MECHANISMS OF ANTIBIOTIC RESISTANCE GENE THROUGH ANIMAL MANURE

Antibiotic-resistant bacteria and viruses are rising due to the widespread dispersal of antibiotic resistance genes in the natural environment. The general use of dung as a fertilizer is one possible cause of this proliferation. Manure contains a variety of resistance genes, increasing the number of environmental resistance genes. While some bacteria are naturally resistant to antibiotics, susceptible bacteria can become resistant through genetic mutations or by acquiring resistance genes (ARGs) from other bacteria through horizontal gene transfer. ARGs are often present in a resistant commensal population, such as in the animal gut, which is selected due to the use of antibiotics (Witte, 2000; Andremont, 2003; Levy and Marshall, 2004). The animal gut microbiome has a vast gene pool that may include numerous antibiotic-resistance factors (Chee-Sanford et al., 2009). Animal faeces, which contain members of the intestinal microbial community and are rich in nutrients, are used as fertilizer in agriculture. This indicates that faeces or manure bacteria can supply resistance-determining agents to the farming environment. Manure may also contain leftover antibiotics and resistance genes (Chee-Sanford et al., 2009; Kim et al., 2011). Farms and the environment around them are more affluent in antibiotics, antibiotic-resistant microorganisms, and ARGs (Campagnolo et al., 2002; Ghosh and LaPara, 2007; Hsu et al., 2014; Zhu et al., 2013). The most crucial way for bacteria to capture ARGs is HGT. One of the following three processes could cause it to include conjugation, transduction, and transformation, as shown in Figure 4.1.

4.2.1 Conjugation

Conjugation transfers genetic material from one cell to another by producing a pilus. Since its discovery in the 1950s, this transfer mechanism has been discovered in various habitats, including water, soil, ocean sediments, activated sludge, and sewage effluent (Davison, 1999). Conjugation is aided by the conjugative machinery, which can be encoded in the chromosome or found on plasmids that can replicate autonomously (Smillie et al., 2010; Wozniak and Waldor, 2010). Plasmids and integrative conjugative elements are frequently included in the transmitted genetic material. The conjugative mechanism can also mobilize non-conjugative plasmids, as evidenced by the IncQ plasmid's extensive host range (Meyer, 2009). Conjugation has been studied more than any other method of HGT (Norman et al., 2009; Guglielmini et al., 2013). Once resistance genes have been established on successful plasmids, they can swiftly spread across strains, species, or even genera, as observed with the blaCTXM ESBL genes, which have transferred to different *Enterobacteriaceae* plasmids (Woerther et al., 2013). Plasmid transfer in pathogens has resulted in the global spread of antibiotic resistance genes (ARGs) for various drug classes, including lactams, quinolones, aminoglycosides, tetracyclines, sulfonamides, and others. The reported spread of plasmids carrying carbapenem resistance across multiple continents (Arcilla et al., 2016), as well as the recent discovery of plasmid-encoded colistin resistance in China (Liu et al., 2016), are cause for concern because they may lead to pan-drug resistance in *Enterobacteriaceae* (Carattoli, 2013). Furthermore, numerous ARGs frequently co-localize on a single plasmid, making spreading multidrug resistance relatively easy.

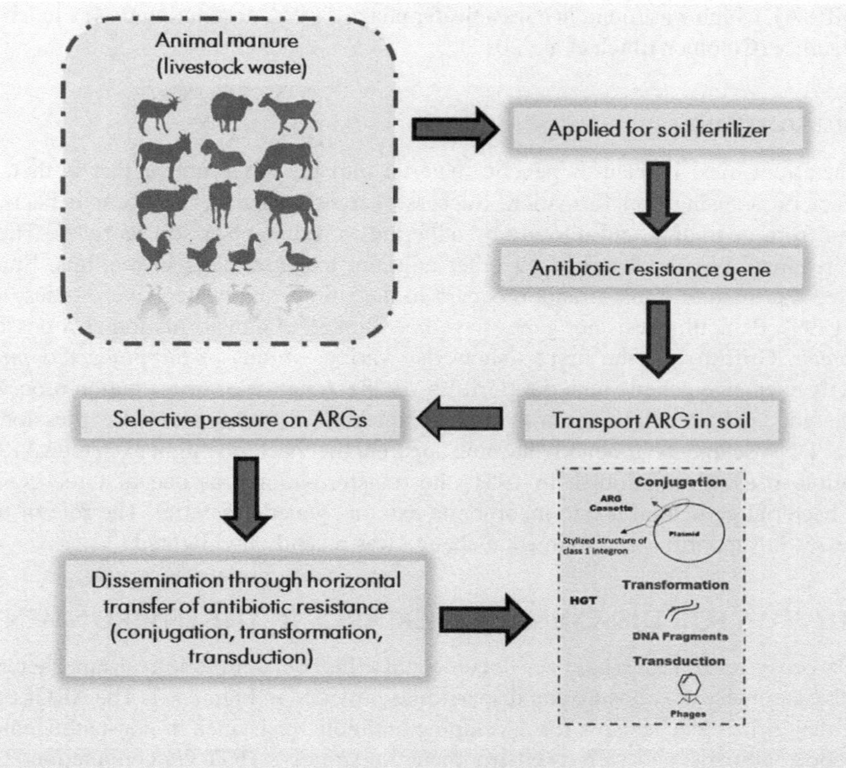

FIGURE 4.1 Pathway of dissemination.

4.2.2 Transduction

During transduction, bacteriophages deliver genetic material to bacteria, making them more suitable than naked DNA for DNA dispersion due to their small size and ability to endure environmental degradation (Muniesa et al., 2013). This process frequently occurs in marine settings, but bacteriophages have also been found in various environments, such as freshwater reservoirs, vegetation, natural lakes, and marine environments, carrying plasmids, chromosome markers, and virulence genes through the environment (Muniesa et al., 2013). The transduction process involves the transfer of DNA between bacteria using bacteriophages or phages, which is more crucial for environmental gene transfer than previously believed (Muniesa et al., 2013). These bacteriophage particles are excellent candidates for facilitating DNA transfer in the ecosystem, as they are more resistant to environmental deterioration than naked DNA, and their condensed size aids in their dissemination (Johnsborg and Havarstein, 2009). Moreover, specific phages are known to infect various bacterial groups, with some having extensive host ranges, making them perfect for transporting genes between ecologically distinct bacterial species and human microbiome communities (Jensen et al., 1998; Muniesa et al., 2013). The transduction mechanism is often observed in marine environments, emphasizing its significance (Jiang and Paul, 1998). Evolutionary studies have revealed that essential prophage origins exist in some bacterial genome segments, underscoring the importance of viral changes to the bacterial chromosome (Brüssow and Hendrix, 2002). Additionally, viral metagenome studies have led to the discovery of β-lactamase genes in activated sludge and urban sewage, indicating the potential for gene transfer through phages (Rolain et al., 2011). One noteworthy example is the mecA gene, responsible for conferring methicillin resistance in methicillin-resistant *S. aureus* bacteria (MRSA), which was found in a wastewater phage's DNA originating from a receiving water treatment facility (Colomer-Lluch et al., 2011).

4.2.3 Transformation

When a bacterium dies, it releases genetic material into the environment that is then absorbed by other bacteria, causing transformation. Successful transformation can occur if bacterial DNA is prevented from degrading and diluting by adhering to sediment or soil particles. The DNA of lysed bacteria may also be absorbed by other adjacent bacterial cells in biofilms. Studies have demonstrated that natural transformation occurs in the soil, groundwater, river bodies, and ocean (Davison, 1999). Penicillin-resistant genes have been linked to transformation as a possible cause of their spread. Griffith was the first to show that various strains of *Streptococcus pneumoniae* could directly exchange genetic material (Griffith, 1928). A natural transformation process appears possible in some bacteria, allowing for the uptake, integration, and functional expression of naked extracellular DNA segments. It quickly became apparent that bacteria might exchange ARGs during transformation to avoid antibiotics. In 1951, the transformation was one of three gene transfer methods a bacterial cell could use to incorporate external genetic material. The role of transformation processes in supporting environmental change was recently highlighted.

4.3 PATHWAYS FOR DISSEMINATION OF ARGS IN THE ENVIRONMENT

Nearly every ecosystem on earth has been documented to harbour antibiotic-resistant bacteria strains. The executive summary of exposure and dispersion is provided in Figure 4.1. The ARGs can spread, and one of the critical mechanisms for developing antibiotic resistance in non-pathogenic bacteria found in people, animals, and various environmental matrices is HGT *via* conjugation. The pathogenic bacteria discharged into the soil via animal manure ultimately become resistant by acquiring resistant genes from the non-pathogenic bacteria present in the soil may catalyze the spread of resistance A possible factor in the formation of ARGs is the presence of high doses of antibiotics in the environment, which promote natural selection. This will further establish the environment, particularly aquatic environments, as a reservoir of antibiotic-resistant genes and their subsequent expansion to dangerous bacteria through direct and indirect genetic transmission channels. Humans are exposed to these ARGs in various ways, allowing them to become a part of their microbiome, as shown in Figure 4.2.

4.4 ARG TRANSMISSION IN THE SOIL-PLANT SYSTEM

Concerns have been expressed concerning the buildup of antibiotic resistance genes (ARGs) in agricultural soils due to the possible transfer of these genes to plant microorganisms and, ultimately, consumers via food intake (Looft et al., 2012). According to research, consuming raw fruits and vegetables may be a mechanism for transmitting ARGs to human microbiomes. ARGs and antibiotic resistance bacteria (ARBs) can be transferred to plant microbiomes via a variety of mechanisms, including phyllosphere microorganism attachment to the leaf surface, invasion and occupation of leaf and root tissue as endophytes, and absorption of antibiotic residues from manure-amended soils. This long-term selection pressure on the plant microbiome may aid in establishing and transmitting resistomes into the human food chain, as shown in Figure 4.3 (Chen et al., 2019).

4.5 APPLICATION OF MANURE ALTERS SOIL'S RESISTOME

In agricultural regions, the resistome is influenced by both human and animal activities (Muurinen et al., 2017). After 30 days of soil application, antibiotics present in manure, such as sulfadiazine, can

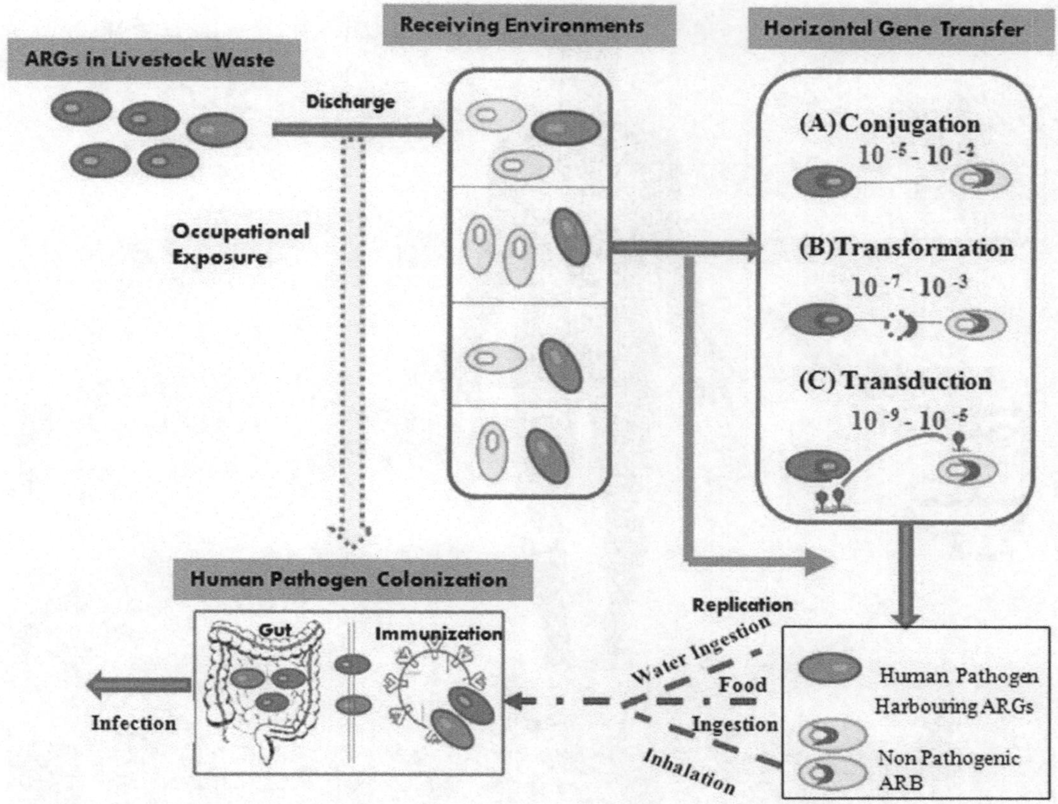

FIGURE 4.2 Transfer pathway of ARGs from livestock waste to human pathogens.

rapidly diminish or even eliminate (Hammesfahr et al., 2008; Kopmann et al., 2013). Consequently, antibiotics and antibiotic resistance genes (ARGs) can be transmitted from soil to human infections, affecting resistance patterns (Gillings, 2013). Environmental bacteria host many ARGs, including symbiotic and harmful types (Lehman et al., 2015). Consequently, bacteria's phylogenetic and taxonomic structure significantly impacts the resistome composition of sludge composts and soils (Peng et al., 2017; Sun et al., 2016). Manure can affect soil microorganisms, chemistry, and morphology (Sun et al., 2016). Rather than the direct import of manure-borne bacteria, it is believed that changes in bacterial composition are responsible for the long-term influence of manure on the soil resistome (Sun et al., 2016; Xun et al., 2016). Using chemical fertilizers also has potentially far-reaching effects on the composition and diversity of soil microbes (Sun et al., 2016; Xun et al., 2016). However, the impact of changes in the bacterial population on the soil ARG resistome in long-term applications of animal manure and chemical fertilizers remains unclear.

In response to manure applications from common farm livestock, the activities of ARGs in agricultural soil varied. The abundance of ARGs increased by a factor of 36 after pig dung was applied, while chicken and cow manure led to increases of 30 and 11, respectively. The impact of different manures on the number of ARGs in agricultural soil follows the order: pig manure > chicken manure > cow manure. This trend corresponds to the influence of different manures on the abundance of 16S rRNA genes in soil, suggesting a connection between the persistence and occurrence of ARGs and soil microorganisms, possibly linked to the variety and survival of host bacteria in the soil (Qian et al., 2018; Zhou et al., 2017).

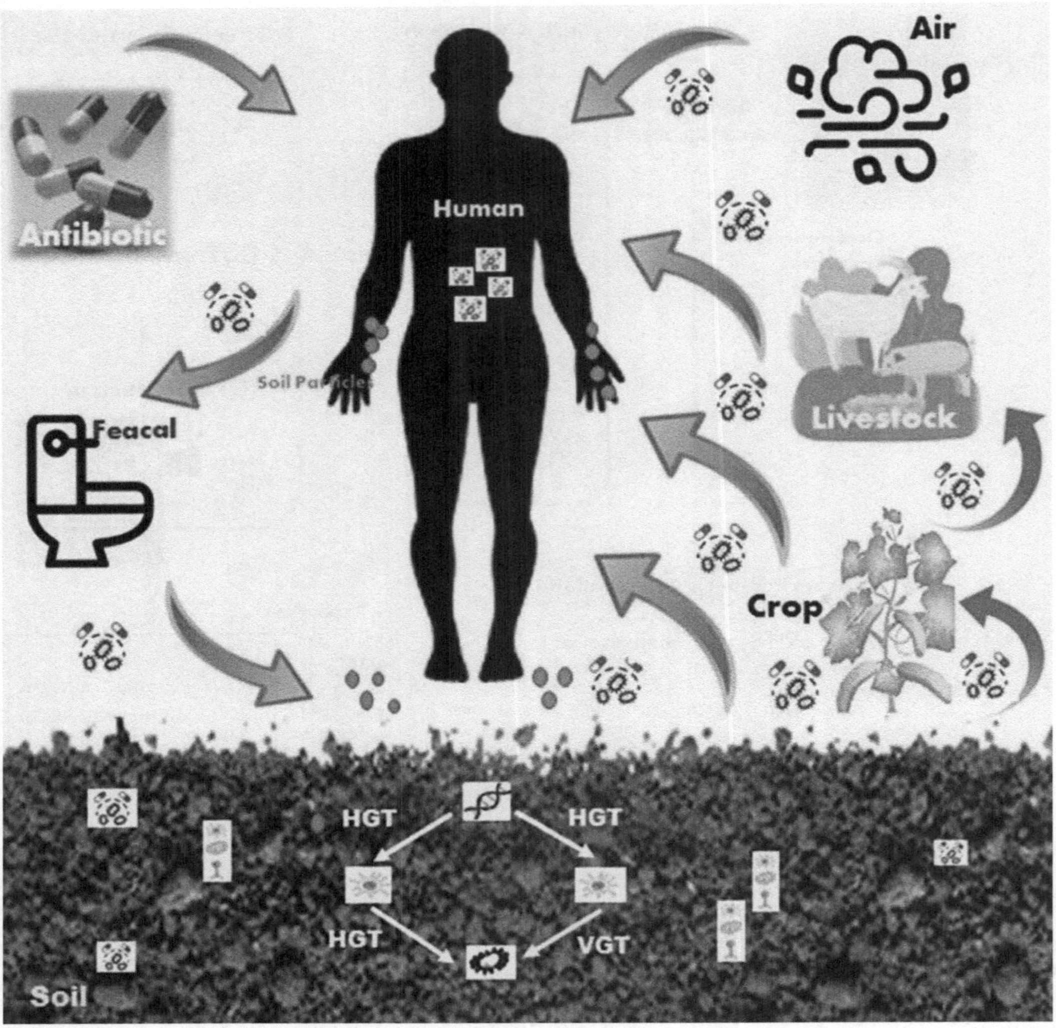

FIGURE 4.3 ARGs transmission in the soil-plant system.

Eating raw leafy vegetables and fruits has been identified as a significant source of antibiotic resistance in humans (Zhang et al., 2011; Verraes et al., 2013). These plants may absorb antibiotics from soil mixed with animal manure, an organic fertilizer widely used in agriculture, potentially stressing the plant microbiome and contributing to antibiotic resistance creation and dissemination in the human food chain. Research has shown that several ARGs can be found in the plant microbiome through HT-qPCR, and the prevalence of ARGs is significantly increased with the use of organic fertilizer, suggesting that plants may acquire exogenous ARGs (Chen et al., 2018). Although the majority of ARG-carrying plant bacteria are non-pathogenic, there is concern about their potential involvement in the horizontal gene transfer (HGT) of ARGs to other bacterial species, including human pathogens, through mobile genetic elements (MGEs) (Rossi et al., 2014). Class I integrons and transposase genes in harvested vegetables indicate that phyllosphere horizontal gene transfer is likely (Wang et al., 2015; Chen et al., 2018). Both the phyllosphere and rhizosphere, where cells cluster and form biofilms, may be key sites for HGT in plant and soil ecosystems, given the high bacterial metabolic rates and MGE mobility in rhizospheres (Van et al., 2003).

4.6 CHANGES IN PLANT RESISTOME OCCURRING DUE TO MANURE APPLICATION

Common soil bacteria can transmit antibiotic-resistance genes to plants through root endophytes, which persist in root systems (Bulgarelli et al., 2013). Some of these endophytes may be related to human diseases, including opportunistic infections, posing a risk to human health (Rosenblueth and Martinez-Romero, 2006). Antibiotic-resistant bacteria associated with soil and manure can infiltrate the plant microbiome through direct root interactions with the soil or spread through the air or motile root endophytes (Zhu et al., 2017). Rainfall, erosion, and manure application all contribute to the spread of pollutants such as antibiotics, antibiotic residues, ARBs, and ARGs, which can then accumulate in the soil and contaminate crops (McEachran et al., 2015). Antibiotic uptake by plants differs from one crop type and growing environment to the next; for example, leafy greens and roots can take up more antibiotics than fruiting grains (Christou et al., 2019). Consuming contaminated vegetables can expose people to bacteria-carrying diseases and infections (Van Hoek et al., 2015).

Adding raw manure from dairy cows to the soil during antibiotic administration, especially with clindamycin, significantly impacts vegetable surfaces' microbiota and resistome composition (Fogler et al., 2019). This research shows that the *tetW* and *sul1* genes are widely distributed and abundant in non-pathogenic bacteria found on people and domestic animals and exhibit high representation in waste streams and low abundance in less-affected environments. These genes may indicate soil contamination with ARBs, ARGs, and antibiotic residues from manure. Using naturally produced lettuce, research by Zhu et al. (2017), found that the number of ARGs was eight times higher than using chemical fertilizers. This research highlights the importance of studying plant resistomes and developing management strategies to prevent the spread of antibiotic resistance in the soil-plant system. ARGs can transfer from manure-amended soils to plants through various pathways, such as attaching to aerial leaf surfaces, surviving as phyllosphere microorganisms, colonizing leaf tissues, or transferring into vegetable roots (Schulz and Boyle, 2006). Earlier research has only focused on certain subsets of the plant microbiome, lacking a holistic perspective on antibiotic resistance transmission. Therefore, a comprehensive study of plant resistomes is crucial for better understanding ARG transmission pathways and including natural resistomes in management strategies and human health risk assessment frameworks (Ashbolt et al., 2013; Koch et al., 2021).

4.7 TECHNOLOGIES FOR REDUCING ARGS IN SOIL

Recent efforts have been made to minimize the spread of antibiotic resistance, especially in manure application. Composting is a sustainable method for removing antibiotic resistance genes (ARGs) from livestock manure, although some ARGs can survive (Qian et al., 2018). However, composting can significantly reduce the abundance of ARGs in manure, although their survival may increase when introduced into the soil (Xu et al., 2019). The high temperature generated during composting can effectively kill ARBs, especially pathogens, and destroy some antibiotics, reducing environmental and human health risks. Enhancing the composting conditions such as temperature, pH, moisture, C/N ratio, and feed amount can further improve the degradation of antibiotics and elimination of ARGs (Cheng et al., 2021; Gurmessa et al., 2020). Converting manure into biochar is more effective than compost amendment in reducing ARGs in soil (Zhou et al., 2019). Reducing ARG intake and controlling their spread in soil, such as through porous materials and bacteriophage treatment, is also crucial. Biochar, eggshell, and biochar-based composites effectively reduce the transfer of ARGs from soil to plants (Chen et al., 2018; Ye et al., 2019) due to the adsorption of heavy metals and antibiotics and changes in the bacterial population linked to ARG profile.

Bacteriophage therapy, a century-old alternative to antibiotics, has gained renewed attention due to the growing threat of antibiotic resistance. Bacteriophages can selectively lyse pathogenic bacteria-carrying ARGs while causing minimal disruption to microbial stability and diversity, making it an eco-friendly strategy for managing or removing the threat of ARGs (Ye et al., 2019).

4.8 TRANSFER OF ARG INTO HUMAN BEINGS

ARG selection, dissemination, and propagation in agroecosystems are highly probable due to the extraordinary diversity of ARGs and MGEs in animal-manured soils. Common soil bacteria can also transfer antibiotic-resistance genes to plants via root endophytes (Bulgarelli et al., 2013). According to Rosenblueth and Martinez-Romero (2006), some endophytes seem to be strongly linked to human diseases, particularly opportunistic ones, and this connection may pose a risk to human health. Antibiotic-resistant bacteria from manure and soil may colonize roots in constant touch with the soil. In contrast, aboveground components, maybe via air particles or root endophyte motility, may also be infected (Zhu et al., 2017). After being excreted by animals, pollutants like antibiotics, antibiotic residues, ARBs, and ARGs may be dispersed throughout the ecosystem by rainfall, erosion, and the use of manure (McEachran et al., 2015). These could accumulate in the soil, increasing the likelihood of positive crop selection and ARB and ARG contamination (Hu et al., 2010). There are several biotic and abiotic variables, as well as crop type, that affect a plant's ability to absorb antibiotics. For instance, fruit and grain have lower contaminant absorption rates than leafy and root vegetables (Christou et al., 2019). People who eat infected vegetables risk contracting illnesses and other infections from germs (Van Hoek et al., 2015). Fogler et al. (2019) found that the microbiota and resistome compositions related to vegetable surfaces were significantly altered by adding raw manure from dairy cows to the soil while administering antibiotics, notably clindamycin. According to their research, the *tetW* and *sul1* genes are found in non-pathogenic bacteria that inhabit humans and domestic animals. According to their findings, the *tetW* and *sul1* genes are widely distributed and abundant in non-pathogenic bacteria that live on humans and domestic animals. It was also shown that these genes have a consistent and highly conserved DNA sequence and are plentiful in waste streams but few in unaffected habitats. As a result, these genes could detect soil contamination from ARB, ARG, and antibiotic residues derived from manure. According to Zhao et al. (2019), this is a widespread issue in which animal manure recycling onto agricultural soils could introduce antibiotic resistance into vegetables, which are frequently consumed raw or after post-processing. Into the previous findings, growing vegetables in soils supplemented with animal manure likely lead to an increase in some ARGs on consumer-ready veggies during harvest or even at the store market (Marti et al., 2013; Rahube et al., 2014; Wang et al., 2015; Zhu et al., 2017; Yang et al., 2018). For example, the abundance of ARGs in organically produced lettuce was eight-fold higher than that of conventionally produced (Zhu et al., 2017).

Direct human contact with ARB can occur by drinking water, eating contaminated food, or even breathing contaminated air. As a result, ARB levels throughout the body (especially the stomach) might rise, making people more susceptible to internal and external infections. Particular attention should be paid to the human digestive system since it serves as the body's primary ARG reservoir and has the largest relative quantity and richness of ARGs observed in both human and non-human environments (Pal et al., 2016). Ingestion of contaminated items (food, trash, residual waste from occupational exposure, or polluted environments) and inhalation of airborne ARBs are two known routes of entry for these compounds into human gut ecosystems (Sanganyado et al., 2019; Ginn et al., 2021). The fate of ingested bacteria, including ARB, is highly species- and strain-dependent and influenced by factors such as the structure of the preexisting microbiome, the presence or absence of drugs, the age of the host, and the context of the meal (Yatsunenko et al., 2012; David et al., 2014). Most ingested bacteria only temporarily colonize the human body under typical nutritional and physiological circumstances, making the variety and density of gut microbiome a major barrier to integrating ingested bacteria (Derrien et al., 2015; Manaia, 2017). Antibiotic treatment or residue ingestion, on the other hand, may dramatically alter the composition of the gut microbiota, leading to more ARB integration through reduced competition and enhanced ARG proliferation. The type and quantity of ARGs in the human gut are affected by residual antibiotics (e.g., >60 ng/kg in meat products). It has also been noted that the changes to the gut flora caused by antibiotics might last for years (Haak et al., 2019). Long-term ARB colonization in the human gut can provide a conducive environment for the development of multi-drug-resistant infections like as methicillin-resistant *Staphylococcus aureus* (MRSA) due to the presence of many factors (Lerner et al., 2017; Krezalek et al., 2018). Strictly anaerobic commensal bacteria keep most ARGs in the human gut and are seldom passed to gut-dwelling opportunistic pathogens (Van Schaik, 2015). Depending on their virulence, ARB invading human gastrointestinal settings can have a significant influence. Livestock-associated MRSA (LA-MRSA) isolates were shown to have lower virulence than either hospital-acquired (HA-MRSA) or community-acquired (CA-MRSA) strains. For instance, compared to CA- or HA-MRSA isolates, LAMRSACC398 is less invasive because it has a reduced ability to attach to human cells. Despite this, LA-MRSA retains a strong cytotoxic potential; different strains are more invasive than others. In addition, it is hypothesized that the colonization of agricultural labourers by LA-MRSA led to its introduction to hospitals and towns where HA- and CA-MRSA formed (Becker et al., 2017). On the other hand, several ARB harboured by livestock may not infect humans due to maladaptation, infection barriers, and immune response(Groeger et al., 2015; Iwasaki et al., 2015). However, such risks remain to be accurately quantified due to insufficient transport and public exposure data and the lack of a clear etiology between resistance observed in livestock waste versus clinical resistance.

4.9 CONCLUSION

Considering the issue of antibiotic misuse and its effect on antibiotic resistance, it is strongly recommended that antibiotic use in livestock production be reduced. Soil manuring, a common practice, has become a concern due to its contribution to the unchecked proliferation of antibiotic-resistant bacteria (ARB) and antibiotic-resistance genes (ARGs) in the ecosystem. Despite efforts to implement better waste and animal management techniques on farms to curb the spread of ARGs, studies have consistently shown the persistence of ARGs in the microbial communities of livestock dung and manured agricultural fields. Over the past thirty years, multidisciplinary research methods have significantly advanced our understanding of how ARGs diffuse from animal waste into the environment. The application of animal waste manure to soil has increased the diversity and abundance of ARGs in manure-amended soils. Although the persistence of ARGs in the soil varies from a few days to several months, the risk remains as bacteria can still take them up through horizontal gene transfer (HGT) mechanisms, potentially reaching humans through the food chain. To effectively tackle this global issue, it is essential to foster collaborative and cross-disciplinary research on the pathways of ARG transmission from animal waste to manured agricultural soil, plants, and the human gastrointestinal environment. The World Health Organization (WHO) has published a comprehensive global action plan on antimicrobial resistance (AMR), adopted as a national action plan in 58 countries. However, debates continue regarding the effectiveness of these guidelines, evident in the lack of full adoption in some regions like India, where only three out of 28 states have converted the national action plan on antibiotic resistance into a state action plan, and the over-the-counter selling of antibiotics remains common. Despite the WHO's Global Action Plan guidelines and the FDA's five-year ban on antibiotics in animal feed, the rise in AMR-related deaths persists, exposing the limitations of current linear approaches.

To effectively combat antibiotic resistance, it is crucial to replace the legacy knowledge of health systems with cross-sectoral and ecologically responsible approaches. Policymakers must review the application and practice of standards to prevent AMR and ensure progress in this battle against antibiotic resistance. Antibiotic misuse poses a severe health concern for the world's living environment and must be urgently addressed to avoid becoming one of the most significant threats to living beings. Livestock producers must be educated about the risks, leading to a pressing need to limit antibiotic usage in animal production. An alarming situation arises from the uncontrolled dispersal of ARGs, ARBs, and AMR through animal waste in the environment. Despite efforts to improve waste and animal management procedures within farms, research has consistently shown the presence of ARGs in the microbial communities of livestock waste and manured farmland. Multidisciplinary research over the past three decades has substantially enhanced our understanding of how ARGs disseminate from animal waste to the environment. The addition of animal waste manure to soil has considerably increased the variety and quantity of ARGs, posing a persistent risk. The durability of ARGs in the soil, ranging from a few days to several months, allows them to be taken up by bacteria through HGT processes, potentially affecting soil microbiota and even pathogenic animal and human bacteria, which may reach humans through the food chain. Collaborative and multidisciplinary research on ARG transmission from animal waste to manured agricultural soil, plants, and human gastrointestinal ecosystems is paramount.

4.10 ALTERNATIVES FOR REDUCING THE RISKS OF ANTIBIOTIC RESISTANCE IN THE FUTURE

Antibiotic resistance is now widely acknowledged as a significant public health problem. It should be recognized that ARGs are a community of ancient genes present in various environmental bacteria that are harmless to people unless transmitted to human pathogens. Consequently, there are still unanswered concerns regarding the relationship between environmental ARGs and human health, and further study is required. In agricultural environments, understanding the transmission channels and processes of ARGs is crucial. Efforts must be made to impede the spread of antibiotic resistance. Therefore, three worldwide strategies aim to minimize antibiotic usage in food animals: banning antibiotics, lowering meat consumption, and charging users for antibiotics in animals. The global prevalence of ARGs in livestock waste and their apparent ability to spread to receiving environments and eventually to humans highlight the need for a better understanding of how to reduce in-farm ARG proliferation (source), mitigate ARG discharge (elimination), and decrease dissemination (fate and transport). This understanding is essential to implement effective measures.

4.11 FIVE CRITICAL TOPICS FOR FUTURE STUDY

Five key areas have been identified for future research to address the problem of antibiotic resistance genes (ARGs) in soil and crops cultivated in soil. The first area involves differentiating between clinically significant ARGs and those that do not confer resistance. Secondly, it is essential to address the problem of antibiotic overuse and misuse in clinics and agriculture by implementing measures to reduce antibiotic consumption. The third aspect is to develop a method for assessing the hazards associated with integrating ARGs and other pollutants. The fourth study area is to understand the extent and circumstances of ARG emergence alongside co-occurring pollutants and their transformation in soil, plants, and biodiversity. Lastly, promoting sustainable green technologies is crucial to mitigating the introduction of ARGs and other contaminants, ensuring a safer and healthier environment.

4.12 ACKNOWLEDGEMENT

We gratefully acknowledge financial support under the Fast Track Scheme of the Department of Science and Technology (DST-SERB), Government of India, New Delhi. The authors are also grateful to the Council of Science and Technology (CST), UP, Center of Excellence (CoE), and the Research and Development Project from the Department of Higher Education, Government of UP for financial support. DK and RG are grateful to University Grants Commission (UGC), New Delhi, for providing them with doctoral fellowships as JRF and SRFs.

REFERENCES

Ahmad, I., Malak, H. A., and Abulreesh, H. H.. 2021. Environmental antimicrobial resistance and its drivers: A potential threat to public health. *J. Glob. Antimicrob. Resist.*27: 101–111.

Andremont, A. 2003. Commensal flora may play key role in spreading antibiotic resistance we need to learn more about commensal flora if we are to better manage this particular window of vulnerability to antibiotic resistance. *ASM News-Ame. Soc. Microbiol.*69(12): 601–607.

Arcilla, M. S., van Hattem, J. M., Matamoros, S., Melles, D. C., Penders, J., de Jong, M. D., and Schultsz, C. 2016. Dissemination of the *mcr-1* colistin resistance gene. *Lancet Infect Dis.*16:147–149.

Ashbolt, N. J., Amézquita, A., Backhaus, T., Borriello, P., Brandt, K. K., Collignon, P., and Topp, E. 2013. Human health risk assessment (HHRA) for environmental development and transfer of antibiotic resistance. *Env. Heal. Perspect* 121(9): 993–1001.

Becker, K., Ballhausen, B., Kahl, B. C., and Köck, R. 2017. The clinical impact of livestock-associated methicillin-resistant Staphylococcus aureus of the clonal complex 398 for humans. *Vet. Microbial.*200: 33–38.

Blau, K., Jacquiod, S.,Sorensen, S. J., Su, J. Q., Zhu, Y. G., Smalla, K., and Jechalke, S. 2019. Manure and doxycycline affect the bacterial community and its resistome in lettuce rhizosphere and bulk soil. *Front. Microbial.*10: 725.

Brüssow, H., and Hendrix, R. W.2002. Phage genomics: Small is beautiful. *Cell* 108 (1): 13–16.

Bulgarelli, D., Schlaeppi, K., Spaepen, S., Ver Loren van Themaat, E., and Schulze-Lefert, P. 2013. Structure and functions of the bacterial microbiota of plants. *Annu. Rev. Plant Biol.*64: 807–838.

Campagnolo, E. R., Kammy, R.,Johnson, A., Karpati, C., Rubin, S., Dana, W., Kolpin, M. T., Meyer, J., and Emilio, E. 2002. Antimicrobial residues in animal waste and water resources proximal to large-scale swine and poultry feeding operations. *Sci. Tot. Environ* 299(1–3): 89–95.

Carattoli, A. 2013. Plasmids and the spread of resistance. *Int. J. Med. Microbiol.*303(6–7): 298–304.

Che, Y., Xia, Y., Liu, L., Li, A. D., Yang, Y., and Zhang, T. 2019. Mobile antibiotic resistome in waste water treatment plants revealed by Nanoporemetagenomic sequencing. *Microbiome*7 (1): 1–13.

Chee-Sanford, J. C.,RoderickI. M., Satoshi, K., Ivan, G. K., Yu, F. L., Anthony, C. Y, Scott, M., and Rustam, I. A. 2009. Fate and transport of antibiotic residues and antibiotic resistance genes following land application of manure waste. *J. Environ. Qual.*38(3): 1086–1108.

Chen, Q. L., Fan, X. T., Zhu, D., An, X. L., Su, J. Q., and Cui, L. 2018. Effect of biocharamendment on the alleviation of antibiotic resistance in soilandphyllosphere of *Brassica chinensis* L. *Soil Biol. Biochem.*119: 74–82.

Chen, Q. L., Cui, H. L., Su, J. Q., Penuelas, J., and Zhu, Y. G. 2019. Antibiotic resistomes in plant microbiomes. *Trends Plant Sci.*24(6): 530–541.

Cheng, D., Liu, Y., Shehata, E., Feng, Y., Lin, H., Xue, J., and Li, Z. 2021. In-feed antibiotic use changed the behaviorsofoxytetracycline, sulfamerazine, and ciprofloxacin and relatedantibiotic resistance genes during swine manure composting. *J. Haz. Mater.*402: 123710.

Christou, A., Papa david, G., Dalias, P., Fotopoulos, V., Michael, C., and Bayona, J. M. 2019. Ranking of crop plants according to their potential to uptake and accumulate contaminants of emerging concern. *Environ. Res.*170: 422–432.

Colomer-Lluch, M., Lejla, I., Juan, J., and Maite, M. 2011. Bacteriophages carrying antibiotic resistance genes in fecal waste from cattle, pigs, and poultry. *J. Antimicrob. Agents.*55(10): 4908–4911.

David, L. A., Maurice, C. F., Carmody, R. N., Gootenberg, D. B., Button, J. E., Wolfe, B. E., Ling, A. V., Devlin, A. S., Varma, Y., Fischbach, M. A., Biddinger, S. B., Dutton, R. J., and Turnbaugh, P. J. 2014. Diet rapidly and reproducibly alters the human gut microbiome. *Nature* 505(7484): 559–563.

Davison, J. 1999. Genetic exchange between bacteria in the environment. *Plasmid.*42(2): 73–91.

Derrien, M., and van HylckamaVlieg, J. E. 2015. Fate, activity, and impact of ingested bacteria within the human gut microbiota. *Trends Microbiol.*23(6): 354–366.

Figueroa-Diva, R. A.,Vasudevan, D., and MacKay, A. A. 2010. Trends in soil sorption coefficients within common antimicrobial families. *Chemosphere* 79(8): 786–793.

Fogler, K., Guron, G. K. P., Wind, L. L., Keenum, I. M., Hession, W. C., and Krometis, L. A. 2019. Microbiota and antibiotic resistome of lettuce leaves and radishes grown in soils receiving manure-based amendments derived from antibiotic-treated cows. *Front. Sustain. Food Syst.*3:22.

Ghosh, S., and Timothy, M. L. P.2007. The effects of subtherapeutic antibiotic use in farm animals on the proliferation and persistence of antibiotic resistance among soil bacteria. *The ISME J.*1(3): 191–203.

Gillings, M. R. 2013. Evolutionary consequences of antibiotic use for the resistome, mobilome and microbial pangenome. *Front. Microbial.*4: 4.

Ginn, O., Berendes, D., Wood, A., Bivins, A., Rocha-Melogno, L., Deshusses, M. A., Tripathi, S. N., Bergin, M. H., and Brown, J. 2021. Open waste canals as potential sources of antimicrobial resistance genes in aerosols in Urban Kanpur, India. *Am. J. Trop. Med. Hyg.*104(5): 1761–1767.

Griffith, F. 1928. The significance of pneumococcaltypes. *J. Hyg. Lond.*27: 113–159.

Groeger, S. E., and Meyle, J. 2015. Epithelial barrier and oral bacterial infection. *Periodontol 2000.*69(1): 46–67.

Guglielmini, J., delaCruz, F., and Rocha, E. P. 2013. Evolution of conjugation and type IV secretion systems. *Mol. Biol. Evol.*30: 315–331.

Gurmessa, B., Pedretti, E. F., Cocco, S., Cardelli, V., and Corti, G. 2020. Manure anaerobic digestion effects and the role of pre- and post-treatments on veterinary antibiotics and antibiotic resistancegenes removal efficiency. *Sci. Tot. Environ.*721: 127532.

Haak, B. W., Lankelma, J. M.,Hugenholtz, F., Belzer, C., de Vos, W. M., and Wiersinga, W. J. 2019. Long-term impact of oral vancomycin, ciprofloxacin and metronidazole on the gut microbiota in healthy humans. *J. Antimicrob. Chemother.*74(3): 782–786.

Hammesfahr, U., Heuer, H., and Manzke, B. 2008. Impact of the antibiotic sulfadiazine and pig manure on the microbial community structure in agricultural soils. *Soil Biol. Biochem.*40: 1583–91.

Hernando-Amado, S., Coque, T. M.,Baquero, F., and Martínez, J. L. 2019. Defining and combating antibiotic resistance from one health and global health perspectives. *Nat. microbial.*4 (9): 1432–1442.

Hsu, J. T., Chia,Y. C., Chu, W. Y., Wei, Lian, G. C., Mao-H, L., Yung-H, L., Chu, M. L., and Chingwen, Y. 2014. Prevalence of sulfonamide-resistant bacteria, resistance genes and integron-associated horizontal gene transfer in natural water bodies and soils adjacent to a swine feedlot in northern Taiwan. *J. Hazard. Mater.*277: 34–43.

Hu, X., Zhou, Q., and Luo, Y. 2010. Occurrence and source analysis of typical veterinary antibiotics in manure, soil, vegetables and groundwater from organic vegetable bases, northern China. *Environ. Pol.*158(9): 2992–2998.

Iwasaki, A., and Medzhitov, R. 2015. Control of adaptive immunity by the innate immune system. *Nat. immunol.*16(4): 343–353.

Jauregi, L.,Lur, E., Itziar, A., and Carlos G. 2021. Antibiotic resistance in agricultural soil and crops associated to the application of cow manure-derived amendments from conventional and organic livestock farms. *Front. Vet. Sci.*8: 633858.

Jensen, E., Holly, C. S. S.,Brenda, R., Thomas, L., Thompson, K. W.,Lee, K, Nickerson, W., and Kokjohn, T. A.1998. Prevalence of broad-host-range lytic bacteriophages of *Sphaerotilusnatans, Escherichia coli*, and *Pseudomonas aeruginosa*. *Appl. Environ. Microbio.*64(2): 575–580.

Jiang, S. C., and John, H. P.1998. Significance of lysogeny in the marine environment: Studies with isolates and a model of lysogenic phage production. *Microb. Ecol.*35(3): 235–243.

Johnsborg, O., and Håvarstein, L. S.2009. Regulation of natural genetic transformation and acquisition of transforming DNA in Streptococcus pneumoniae. *FEMS microbial. Rev.*33(3): 627–642.

Karkman, A., Do, T. T., Walsh, F., and Virta, M. P. J. 2018. Antibiotic-resistance genes in waste water. Antibiotic-resistance genes in waste water. *Trends Microbiol* 26(3): 220–228.

Keenum, I., Robert K., Williams, P. R., Garner, E., Katharine, D., Knowlton, F., and Pruden, A. 2021. Combined effects of composting and antibiotic administration on cattle manure–borne antibiotic resistance genes. *Microbiome*9 (1): 1–16.

Kim, K. R.,Owens, G., Kwon, S. I., So, K. H.,Lee, D. B., and Ok, Y. S. 2011. Occurrence and environmental fate of veterinary antibiotics in the terrestrial environment. *Water Air Soil Pollut.*214(1): 163–174.

Koch, N., Islam, N.F., Sonowal, S., Prasad, R., Sarma, H. (2021). Environmental antibiotics and resistance genes as emerging contaminants: Methods of detection and bioremediation. *Current Research in Microbial Sciences* 2, 100027. https://doi.org/10.1016/j.crmicr.2021.100027

Kopmann, C., Jechalke, S., and Rosendahl, I. 2013. Abudance and transferability of antibiotic resistance as related to the fate of sulfadiazine in maize rhizosphere and bulk soil. *FEMS Microbiol. Eco.*83: 125–34.

Krezalek, M. A.,Hyoju, S., Zaborin, A., Okafor, E., Chandrasekar, L., Bindokas, V., and Alverdy, J. C.2018. Can methicillin-resistant Staphylococcus aureus silently travel from the gut to the wound and cause post-operative infection? Modeling the "Trojan Horse Hypothesis". *Ann. Surg* 267(4): 749–758.

Kümmerer, K. 2004. Resistance in the environment. *J. Antimicrob. Chemo* 54(2): 311–320.

Lehman, R. M., Cambardella, C. A., Stott, D. E., Acosta-Martinez, V., Manter, D. K., Buyer, J. S., Maul, J. E., Smith, J. L., Collins, H. P., Halvorson, J. J., and Kremer, R. J. 2015. Understanding and enhancing soil biological health: The solution for reversing soil degradation. *Sustainability* 7(1): 988–1027.

Lerner, A., Matthias, T., and Aminov, R. 2017. Potential effects of horizontal gene exchange in the human gut. *Front. Immunol.*8: 1630.

Levy, S. B., and Marshall, B. 2004. Antibacterial resistance worldwide: Causes, challenges and responses. *Nat. Med.*10: 122–129.

Liu, Y. Y.,Wang, Y.,Walsh, T. R.,Yi, L. X., Zhang, R., Spencer, J., Doi, Y., Tian, G., Dong, B., Huang, X., and Yu, L. F.2016. Emergence of plasmid-mediated colistin resistance mechanism MCR- 1 in animals and human beings in China:A microbiological and molecular biological study. *Lanc. Infect. Dis.*16:161–168.

Looft, T., Johnson, T. A., Allen, H. K., Bayles, D. O., Alt, D. P., Stedtfeld, R. D., and Stanton, T. B. 2012. In-feed antibiotic effects on the swine intestinal microbiome. *Proc. Natl. Acad. Sci. U.S.A.*109(5): 1691–1696.

Manaia, C. M. 2017. Assessing the risk of antibiotic resistance transmission from the environment to humans: Non-Direct proportionality between abundance and risk. *Trends Microbiol.*25(3): 173–181.

Manyi-Loh, C. E.,Okoh, A. I., and Lues, R. 2023. Prevalence of multidrug-resistant bacteria (Enteropathogens) recovered from a blend of pig manure and pinewood saw dust during anaerobic co-digestion in a steel biodigester. *Int. J. Environ. Res. Public Health.*20 (2): 984.

Manyi-Loh, C., Mamphweli, S., Meyer, E., and Okoh, A. 2018. Antibiotic use in agriculture and its consequential resistance in environmental sources: Potential public health implications. *Molecules* 23(4): 795.

Marti, R., Scott, A., Tien, Y. C., Murray, R., Sabourin, L., Zhang, Y., and Topp, E. 2013. Impact of manure fertilization on the abundance of antibiotic-resistant bacteria and frequency of detection of antibiotic resistance genes in soil and on vegetables at harvest. *Appl. Environ. Microbial.*79 (18): 5701–5709.

McEachran, A. D., Blackwell, B. R., Hanson, J. D., Wooten, K. J., Mayer, G. D., and Cox, S. B. 2015. Antibiotics, bacteria, and antibiotic resistance genes: Aerial transport from cattle feed yards via particulate matter. *Environ. Health Perspect.*123: 337–343.

Meyer, R. 2009. Replication and conjugative mobilization of broad host-range IncQplasmids. *Plasmid.*62(2): 57–70.

Muniesa, M., Colomer-Lluch, M., and Jofre, J. 2013. Potential impact of environmental bacteriophages in spreading antibiotic resistance genes. *Fut. Microbial.*8(6): 739–751.

Muurinen, J., Stedtfeld, R., Karkman, A., Parnanen, K., Tiedje, J., and Virta, M. 2017. Influence of manure application on the environmental resistome under Finnish agricultural practice with restricted antibiotic use. *Environ.Sci.technol.*51(11): 5989–5999.

Norman, A. L., Hansen H., and Soren, J. S. 2009. Conjugative plasmids: Vessels of the communal gene pool. *Philos. Trans. R. Soc. Lond. Biol. Sci.*364(1527): 2275–2289.

O'neill, J. I. M. 2014. Antimicrobial resistance: Tackling a crisis for the health and wealth of nations. *Rev. Antimicrob. Resist.*1: 1–20.

Pal, C., Bengtsson-Palme, J., Kristiansson, E., and Larsson, D. G.2016. The structure and diversity of human, animal and environmental resistomes. *Microbiome* 4(1): 54.

Peng, S., Feng, Y., Wang, Y., Guo, X., Chu, H., and Lin, X., 2017. Prevalence of antibiotic resistance genes in soils after continually applied with different manure for 30 years. *J. Hazard. Mater.*340: 16–25.

Pérez-Valera, E., Kyselkova, M., Ahmed, E., Sladecek, F. X. J., Goberna, M., and Elhottova, D. 2019. Native soilmicroorganisms hinder the soil enrichment with antibiotic resistance genes following manure applications. *Sci. Rep.*9: 1–10.

Qian, X., Gu, J., Sun, W., Wang, X. J., Su, J. Q., and Stedfeld, R. 2018. Diversity, abundanceand persistence of antibiotic resistance genes in various types of animal manure following industrial composting. *J. Hazard. Mater.*344: 716–722.

Rahube, T. O., Marti, R., Scott, A., Tien, Y. C., Murray, R., Sabourin, L., Zhang, Y., Duenk, P., Lapen, D. R., and Topp, E., 2014. Impact of fertilizing with raw or anaerobically digested sewage sludge on the abundance of antibiotic-resistant coliforms, antibiotic resistance genes, and pathogenic bacteria in soil and on vegetables at harvest. *Appl. Environ. Microbiol.*80 (22): 6898–6907.

Ranjalkar, J., and Chandy, S. J.2019. India's National action plan for antimicrobial resistance – An overview of the context, status, and way ahead. *Fam. Med. Prim. Care Rev.*8(6): 1828–1834.

Rolain, J. M.,Laura, F., Christelle, D., and Didier, R. 2011. Bacteriophages as vehicles of the resistome in cystic fibrosis. *J. Antimicrob. Chemother.*66(11): 2444–2447.

Rosenblueth, M., and Martínez-Romero, E. 2006. Bacterial endophytes and their interactions with hosts. *Mol. Plant Microbe Interact.* 19(8): 827–837.

Rossi, F. 2014. Horizontal gene transfer among microorganisms in food: Current knowledge and future perspectives. *Food Microbiol.*42:232–243.

Safdar, M., Ozaslan, M., Khailany, R. A., Latif, S., Junejo, Y., Saeed, M., Al-Attar, M. S., and Kanabe, B.O. 2020. Synthesis, characterization and applications of a novel platinum-based nanoparticles: Catalytic, antibacterial and cytotoxic studies. *J. Inorg. Organomet. Poly. Mater.*30: 2430–2439.

Sanganyado, E., and Gwenzi, W. 2019. Antibiotic resistance in drinking water systems: Occurrence, removal, and human health risks. *Sci. Tot. Environ.*669: 785–797.

Schulz, B., and Boyle, C. 2006. What are endophytes?. In *Microbial Root Endophytes,* ed. B. J. E. Schulz, C. J. C. Boyle, and T. N.Seiber, 1–13. Berlin Heidelberg: Springer Berlin Heidelberg.

Sharma, V. K., Johnson, N., Cizmas, L., McDonald, T. J., and Kim, H. 2016. A review of the influence of treatment strategies on antibiotic resistant bacteria and antibiotic resistance genes. *Chemosphere* 150: 702–714.

Smillie, C., Garcillán-Barcia, M. P., Francia, M. V., Rocha, E. P., and De La Cruz, F. 2010. Mobility of plasmids. *Microbiol. Mol.Biol. Rev.*74(3): 434–452.

Sun, D., Lan, Y., Xu, E. G., Meng, J., and Chen, W. 2016. Biochar as a novel niche for culturing microbial communities in composting. *Was. Managem.*54: 93–100.

Van Boeckel, T. P., Glennon, E. E., Chen, D., Gilbert, M., Robinson, T. P., Grenfell, B. T., Levin, S. A., Bonhoeffer, S., and Laxminarayan, R. 2017. Reducing antimicrobial use in food animals. *Science* 357(6358): 1350–1352.

Van Boeckel, T. P., Pires, J., Silvester, R., Zhao, C., Song, J., Criscuolo, N. G., and Laxminarayan, R. 2019. Global trends in antimicrobial resistance in animals in low-and middle-income countries. *Science* 365(6459): 1944.

Van E. J.D., Turner, S., and Bailey, M.J. 2003. Horizontal gene transfer in the phytosphere. *New Phytol.* 157(3), 525–537.

Van Elsas, J. D. 2003. Horizontal gene transfer in the phytosphere. *New Phytol.*157: 525–537.

Van Hoek, A.H., Veenman, C., van Overbeek, W. M., Lynch, G., de RodaHusman, A. M., and Blaak, H. 2015. Prevalence and characterization of ESBL-and AmpC-producing Enterobacteriaceae on retail vegetables. *Int. J. Food Microbial.*204: 1–8.

Van Schaik, W. 2015. The human gut resistome. *Philos. Trans. R. Soc. Lond., B, Biol. Sci.*370(1670): 20140087.

Verraes, C. 2013. Antimicrobial resistance in the food chain: A review. *Int. J. Environ. Res. Pub. Heal.*10: 2643–2669.

Vinayamohan, P. G., Pellissery, A. J., and Venkitanarayanan, K. 2022. Role of horizontal gene transfer in the dissemination of antimicrobial resistance in food animal production. *Curr. Opin. Food Sci.*47: 100882.

Wang, F. H.,Qiao, M., Chen, Z., Su, J. Q., and Zhu, Y. G. 2015. Antibiotic resistance genes in manure-amended soil and vegetables at harvest. *J. hazard. Mater.*299: 215–221.

Wegst-Uhrich, S. R., Navarro, D. A., Zimmerman, L., and Aga, D. S.2014. Assessing antibiotic sorption in soil: A literature review and new case studies on sulfonamides and macrolides. *Chem Centr J.*8: 1–12.

Witte, W. 2000. Ecological impact of antibiotic use in animals on different complex microflora: environment. *Int. J. Antimicrob. Agents.*14(4): 321–325.

Woerther, P. L., Burdet, C., Chachaty, E., and Andremont, A. 2013. Trends in human fecal carriage of extended-spectrum β-lactamases in the community: Toward the globalization of CTX-M. *Clinic.microbiol.rev.* 26(4):744–758.

Wozniak, R. A., and Waldor, M. K.2010. Integrative and conjugative elements: Mosaic mobile genetic elements enabling dynamic lateral gene flow. *Nat. Rev. Microbiol.*8(8): 552–563.

Wu, J., Wang, J., Li, Z., Guo, S., Li, K., Xu, P., Ok, Y. S.,Jones, D. L., and Zou, J. 2023. Antibiotics and antibiotic resistance genes in agricultural soils: A systematic analysis. *Crit Rev. Environ. Sci. Technol.*53 (7): 847–864.

Xiao, Y., and Li, L. 2016. China's national plan to combat antimicrobial resistance. *Lancet Infect. Dis.*16(11): 1216–1218.

Xu, M., Stedtfeld, R. D., Wang, F., Hashsham, S. A., Song, Y., Chuang, Y., Fan, J., Li, H., Jiang, X., and Tiedje, J. M. 2019. Composting increased persistence of manure-borne antibiotic resistance genes in soils with different fertilization history. *Sci. Tot. Environ.*689: 1172–1180.

Xun, W., Zhao, J., Xue, C., Zhang, G., Ran, W., Wang, B., Shen, Q., and Zhang, R. 2016. Significant alteration of soil bacterial communities and organic carbon decomposition by different long-term fertilization management conditions of extremely low-productivity arable soil in South China. *Environ. Microbial.*18(6): 1907–1917.

Yang, L., Liu, W., Zhu, D., Hou, J., Ma, T., Wu, L., and Christie, P. 2018. Application of biosolids drives the diversity of antibiotic resistance genes in soil and lettuce at harvest. *Soil Biol. Biochem.*122: 131–140.

Yatsunenko, T., Rey, F. E., Manary, M. J., Trehan, I., Dominguez-Bello, M. G., Contreras, M., Magris, M., Hidalgo, G., Baldassano, R. N., Anokhin, A. P., and Heath, A. C. 2012. Human gutmicrobiome viewed across age and geography. *Nature* 486(7402): 222–227.

Ye, Y., Ngo, H. H., Guo, W., Chang, S. W., Nguyen, D. D., Liu, Y., Nghiem, L. D., Zhang, X., and Wang, J. 2019. Effect of organic loading rate on the recovery of nutrients and energy in a dual-chamber microbial fuel cell. *Biores. Technol.*281: 367–373.

Zhang, D. D.,Lin, L. F., Luo, Z. X., Yan, C. Z., Zhang, X. 2011. Occurrence of selected antibiotics in Jiulongjiang river in various seasons, South China. *J. Environ. Monit.*13: 1953–1960.

Zhao, X., Wang, J., Zhu, L., and Wang, J. 2019. Field-based evidence for enrichment of antibiotic resistance genes and mobile genetic elements in manure-amended vegetable soils. *Sci. Tot. Environ.*654: 906–913.

Zhou, X., Qiao, M., and Wang, F. H. 2017. Use of commercial organic fertilizer increases the abundance of antibiotic resistance genes and antibiotics in soil. *Environ. Sci. Pollut. Res.*24 (1): 701–710.

Zhou, X., Qiao, M., Su, J. Q., Wang, Y., Cao, Z. H., Cheng, W. D., and Zhu, Y. G. 2019. Turning pig manure into biochar can effectively mitigate antibiotic resistance genes as organic fertilizer. *Sci. Tot. Environ.*649: 902–908.

Zhu, B., Chen, Q., Chen, S., and Zhu, Y. G.2017. Does organically produced lettuce harbor higher abundance of antibiotic resistance genes than conventionally produced?. *Environ. Internat.98*: 152–159.

Zhu, Y. G., Johnson, T. A., Su, J. Q., Qiao, M.,Guo, G. X., Stedtfeld, R. D.,Hashsham, S. A., and Tiedje, J. M. 2013. Diverse and abundant antibiotic resistance genes in Chinese swine farms. *Proc. Natl. Acad. Sci. U.S.A.*110(9): 3435–3440.

5 Antibiotic Residues in Freshwater

Antimicrobial Resistance Development and Bioremediation

Vaishnavi Shedge, Shefali Gupta, Nidhi V. Maheshwari, and Seema Sambrani

5.1 INTRODUCTION

Microbes are ubiquitous in the air we breathe, the soil we walk on, the water we drink, and even on and in our bodies. While some microbes are beneficial, others can be quite harmful. These harmful microbes are called pathogens and are capable of causing an infection. Pathogens like bacteria, fungi, or viruses can cause infections when they come into contact with susceptible individuals. Antibiotics were first introduced into the medical system for treating infections caused by various pathogens. During the period known as the "golden age of antibiotic discoveries", a wide variety of antibiotics were found and subsequently grouped into distinct classes. In later years, antibiotics were utilized not only for treating human patients but also for animals and plants. Antibiotics have established themselves as an essential component of contemporary medicine and have been extensively used in surgeries, organ transplants, chemotherapy, etc.

The US Food and Drug Administration (FDA) defines antibiotics as drugs that can treat infections brought on by bacteria only and not viruses. Antibiotics are typically grouped into two categories viz broad-spectrum antibiotics and narrow-spectrum antibiotics depending on the ability to treat different types of bacteria. Broad-spectrum antibiotics are effective against various Gram-positive and Gram-negative bacteria. On the contrary, narrow-spectrum antibiotics are drugs that are effective against selective bacterial species.

Antibiotic resistance is a problem that has arisen among bacterial populations as a result of both inappropriate and excessive antibiotic use. Antibiotic resistance has been on the rise in recent years and thus it is becoming difficult to treat infections. The current scenario is worrisome due to the fact that antibiotics are no longer effective against disease-causing organisms. In addition to this, the antibiotics used are making their way into the water system through a variety of different pathways. These antibiotics and antibiotic residues are not only harmful to the environment but also to the natural inhabitants residing within.

5.2 APPLICATIONS OF ANTIBIOTICS

Antibiotics are categorized based on their biological activity spectrum, chemical composition, and the mechanism by which they combat infection. Antibiotics can be categorized into a variety of different groups, including β-lactam, aminoglycosides, chloramphenicol, glycopeptides, annamycin, streptogramins, sulfonamides, tetracyclines, macrolides, oxazolidinones, quinolones, and lipopeptides. Having a wide range of potential applications such antibiotics are extensively used.

DOI: 10.1201/9781003268895-5

5.2.1 Healthcare System

Disease-causing microorganisms are capable of invading human tissue and have a negative impact on the processes that occur within cells. Damage to the cells of the host can be caused both by the microorganisms themselves and by the toxins that they produce. Antibiotics are time-tested; life-savers that have been put to use to treat such infections or diseases. Antibiotics can either stop the growth of microorganisms or simply kill them, which allows them to control the damage that is caused by infectious agents. Antibiotics disable the ability of bacteria to replicate and spread infection. Antibiotics have been shown to be effective against both acute infections and chronic infections, as well as in the prevention of infections. Antibiotics combat infections by attacking bacterial systems such as cell membranes, bacterial cell walls or other synthesis processes of nucleic acid, protein or biological compounds. Antibiotics have been proven to be a cornerstone of modern medicine. WHO-adapted antibiotics such as fluoroquinolones, cefixime, azithromycin, ceftriaxone, cefotaxime, etc. are being used for treating typhoid fever. In recent years, antibiotics have found extensive applications, particularly in the treatment of sepsis, as well as surgeries, chronic conditions like diabetes, organ transplants, dialysis, and cancer care. It has been discovered that antibiotics can promote cancer cell apoptosis. Additionally, antibiotics can slow the growth of cancer cells and prevent them from spreading to other parts of the body. Antibiotics are increasingly being utilized as a therapeutic aid in the treatment of a wide variety of cancers (Kırmusaoğlu et al. 2019, Gao et al. 2020, Bhutta 2006).

5.2.2 Food Industry

Animals that are raised for human consumption are susceptible to infections in the same way that humans are. Animals that are raised or bred specifically for the purpose of producing meat or other foods for human consumption are referred to as food animals. Beef cattle, dairy cattle, goats, sheep, deer, pigs, broiler chickens, layer chickens, and ducks are bred specifically for the purpose of producing food products such as eggs, milk, and meat. Infectious diseases, including respiratory infections, gastrointestinal infections, reproductive infections, and other bacterial infections, are common in animals raised for food. Antibiotics have primarily two applications for treating infectious diseases that can affect animals: prevention and treatment. In addition to this, antibiotics act as growth promoters for improving feed conversion efficiency, improving carcass quality, and increasing economic production. Antibiotics used in veterinary medicine include β-lactams (such as cephalosporin and penicillin), macrocyclines (such as ansamycins, aminoglycosides and glycopeptides), macrolides, tetracyclines, chloramphenicol, spectinomycin, sulfonamides, lincosamide, nitrofurans, nitroimidazoles, polymyxin, and trimethoprim. Antibiotics were also utilized in the process of extending the shelf life of processed foods (Chausali et al., 2022; Karlo et al., 2023). Antibiotics like chlortetracycline, oxytetracycline, nisin, and nystatin were utilized in the process of food preservation for poultry and fish (Sivaraman and Parady 2018, Arsène et al. 2022).

5.2.3 Veterinary Medicine

Antibiotics are an essential tool for the prevention of disease in animals, as well as their control and treatment. Antibiotics are versatile medicines that can be used to treat a clinical illness in a single animal or in a large number of animals. Infections to the skin, wounds, otitis externa, and urinary and respiratory tracts are commonly seen among animals that need to be treated using antibiotics. Antimicrobial classes such as penicillin, macrolides, cephalosporins, chloramphenicol, lincosamides, tetracyclines, potentiated sulphonamides, aminoglycosides, fusidic acid, and fluoroquinolones are used frequently in the treatment of infectious diseases in veterinary medicine of small animals (Guardabassi 2004).

5.2.4 Controlling Plant Disease

Heavy agricultural losses are observed worldwide due to infections caused by plant pathogens. Also, the demand for food is increasing worldwide due to the growing population. In order to meet the increasing demand, agricultural loss due to bacterial infections must be controlled. Out of the total global food production, 10% is lost to plant diseases and infections. The most common pathogens seen amongst plants are parasitic plants, bacteria, fungi, viruses, oomycetes, and nematodes. Some of the disease-causing pathogens in plants are *Agrobacterium*, *Pectobacterium*, *Erwinia*, *Xanthomonas*, *Pseudomonas*, *Xylella*, *Ralstonia*, and *Dickeya*. Antibiotics were introduced initially to control infection in high-value ornamental plants, fruit, crops, and vegetables. The antibiotics used in plant agriculture include streptomycin, gentamicin, oxytetracycline, and oxolinic acid. These antibiotics were generally used to treat the blight of apples, pears, and other plants (McManus et al. 2002).

5.3 ANTIBIOTICS ENTERING THE WATER SYSTEM

Water ecosystems are a crucial component of the environment because they provide food, water, and niches for a variety of habitats, including human habitats. But new pollutants like antibiotic residues have contaminated these aquatic ecosystems. Worldwide, natural water sources have been found to contain antibiotic residues, and these residues are becoming more concentrated because of the widespread use of antibiotics. There are several ways that antibiotics enter natural water resources, including effluents from wastewater treatment facilities, pharmaceutical industries, and animal and aquaculture effluents (Koch et al., 2021).

5.3.1 Effluents from Wastewater Treatment Plants

During their daily operations, wastewater treatment plants (WWTP) deal with a wide variety of pollutants. They are constructed in such a way that it is not possible to completely remove all microorganisms from the environment. The design of WWTPs places a primary emphasis on reducing organic matter, nitrogen, and phosphorus compounds – compounds that are accountable for eutrophication and the reduction of oxygen in water bodies – in order to improve water quality. Because of this, pollutants such as antibiotics, antibiotic residues, and bacteria resistant to antibiotics are not treated properly and are allowed to remain in the environment. The streams of wastewater that have been partially treated are discharged into bodies of water. Microorganisms and organic compounds in high concentrations are both contributing to the spread of antibiotic resistance genes among bacterial populations. Excretion of antibiotics from the body through urine and faeces is a significant antibiotic polluter in the environment. The antibiotic or antibiotic residues do not degrade completely or only partially into the sludge. If the effluents from these WWTPs are not properly treated they can act as a major carrier for antibiotics, which can lead to the contamination of freshwater sources (Osińska et al. 2020, Danner 2019).

5.3.2 Pharmaceutical Industries

Antibiotics from the pharmaceutical and cosmetics industries are considered industrial sources. The pharmaceutical industries that produce antibiotics as a product are significant contributors to the contamination that occurs in the natural environment. The effluents from pharmaceutical plants, which contain antibiotics, are discharged into freshwater without first undergoing the appropriate treatment. Recent research reports from Asian countries state that they are present in higher concentrations than hospitals or aquaculture sites (Sosa-Hernandez et al. 2021). Therefore, untreated effluents from pharmaceutical manufacturing facilities are a source of antibiotic pollution in freshwater environments.

5.3.3 Animal Husbandry and Aquaculture

Antibiotics have been found in animal manure too. The antibiotics or antibiotic residues seep through the pile into the surface or groundwater. This phenomenon is frequently observed for antibiotics that have a high affinity for water or those that can be dissolved in water. Freshwater can be contaminated by various water sources, including wastewater from irrigation, runoff caused by accidents, and raw animal manure. Additionally, the unintentional release of feed or accidental spills results in the release of antibiotics in freshwater (Manyi-Loh et al. 2018).

Antibiotics are utilized to a significant degree in fish farming on an industrial scale to both satisfy the ever-increasing consumer demand as well as to ensure the good health of the fish. In order to encourage the fish to develop to its full potential, antibiotics are frequently included in specially formulated animal feed. Thus, antibiotics are introduced to freshwater through aquaculture too. These antibiotics, which are typically administered as a routine part of aquaculture, are typically either not consumed or not absorbed. Additionally, the inability of fish to properly metabolize antibiotics contributes further to the rise in antibiotic concentrations found in freshwater environments (Silva et al. 2021).

5.4 ANTIBIOTIC RESISTANCE

Antibiotic resistance is a growing concern across the globe. Antibiotic resistance is usually seen in bacteria. The bacteria evolve using different mechanisms to evade the effect of antibiotics which are designed to kill them. The inappropriate use of antibiotics is a significant contributing factor to antibiotic resistance. Natural factors such as selective pressure, mutations, gene transfer, and societal factors such as inappropriate or overuse, inadequate diagnosis, and unprocessed disposal have led to antibiotic resistance in bacteria (Prasad et al., 2020). Figure 5.1 describes the uses of antibiotics in various sectors, the route of entrance into water bodies, and mechanisms and factors for antibiotic resistance.

5.4.1 Selective Pressure

Bacteria are able to become better adapted to the environment in which they live as a result of a process known as natural selection. The gradual development of antibiotic resistance is a typical process by which bacteria gradually adapt to the environments in which they are found. Antimicrobial agents apply selective pressure on bacterial populations, which, in turn, speeds up the evolution of antibiotic resistance in those populations. When there is antibiotic selection pressure in the environment, random mismatch errors that occur during the process of DNA replication and repair have the potential to lead to the emergence of mutations. Mutations cause changes in the genetic makeup of an organism that is resistant to antibiotics. This is one of the ways in which genetic variability can express itself in bacterial populations, and it provides the raw material that is necessary for the development of antibiotic resistance. There will always be a subset of individuals within any large population that are carriers of *de novo* mutations or genes acquired through horizontal gene transfer as a direct result of the environments to which they are exposed. Antibiotic resistance wipes out bacterial strains that do not have resistant mutations, while simultaneously selecting bacterial strains having resistant genes. As a result of their increased likelihood of survival, resistant strains continue to reproduce and eventually come to dominate the population (Hasan 2021).

5.4.2 Mutation

The susceptible bacterial population develops mutated genes that affect drug activity. As a result of these mutations, the cells are able to survive the presence of the antimicrobial molecule. The antibiotics eliminate the susceptible population once a mutant strain becomes resistant to it, leaving

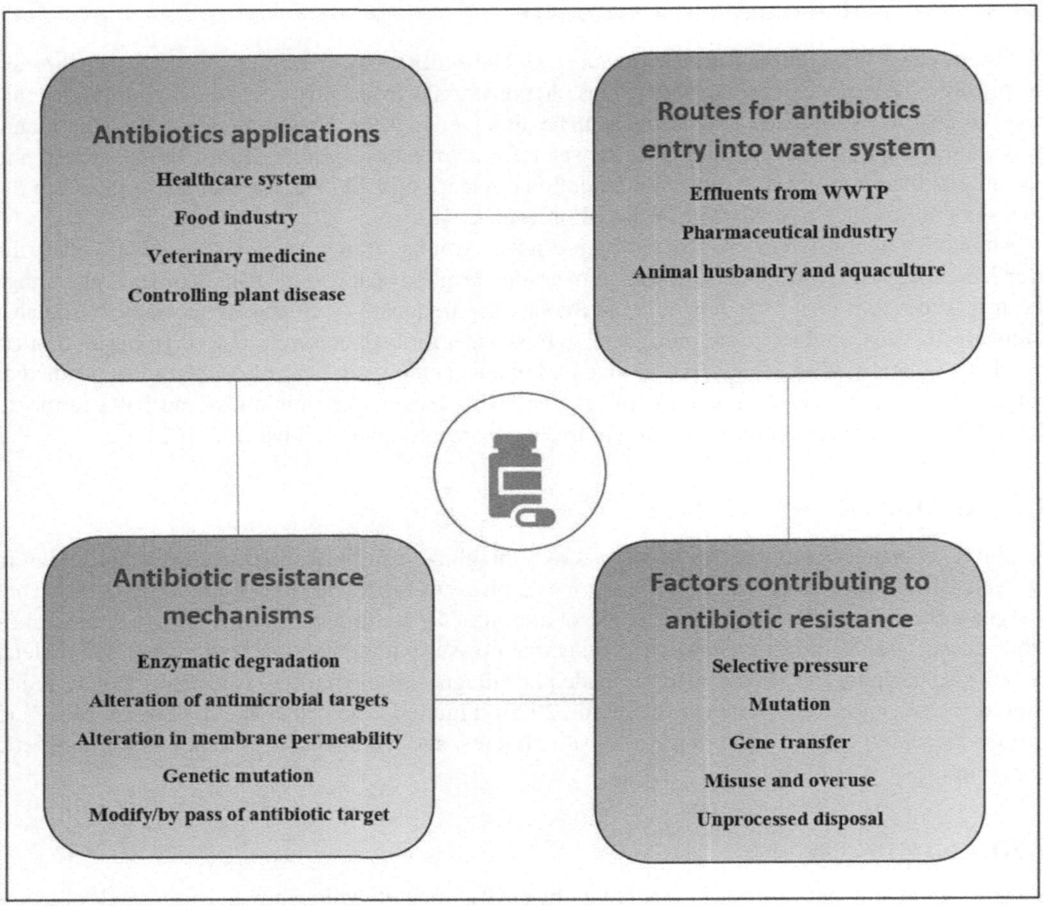

FIGURE 5.1 A framework of antibiotics uses and antibiotic resistance.

behind only those bacteria that are resistant. Mutational changes leading to resistance cause a decrease in bacterial fitness and are maintained in the presence of antibiotics, if necessary.

There are various mechanisms followed by bacteria to acquire antibiotic resistance:

(i) Antimicrobial target modification.
(ii) Extrude harmful molecules by activation of efflux mechanism.
(iii) Decrease in intake of drug.
(iv) Modulating regulatory networks by making changes in metabolic pathways.

Therefore, resistance that arises from acquired mutational alterations is multifaceted and the complexity degree varies (Munita and Arias 2016).

5.4.3 Gene Transfer

Acquiring foreign DNA material by bacteria through horizontal gene transfer (HGT) is an important driver for the evolution of bacteria. HGT in bacteria is the leading cause of the development of antibiotic resistance. Intrinsic genetic determinants of resistance are harboured by bacteria that share an antibiotic-containing environment. In addition, the "environmental resistome" is an abundant source of clinically relevant antibiotic resistance genes in bacteria. Furthermore, reports state the link between the increase in antibiotic resistance to the excessive use of antibiotics to treat infections.

In the traditional model, bacteria acquire genetic material from their environment through one of three primary processes: transformation, also known as the integration of naked DNA; conjugation; and transduction, which is phage-mediated. Of these, transformation is the simplest HGT, whereas conjugation is the most efficient one. The valuable genetic information uses mobile genetic elements during the process of conjugation. Plasmids and transposons are the most important mobile genetic elements because of their roles. Integrons are site-specific recombination systems that recruit open frames in mobile gene cassettes. Thus, integrons are a major source for the accumulation of antimicrobial resistance genes. The robust strategy of integrons for genetic interchange makes them the prime cause of bacterial evolution. Integrons adding new genes into bacterial chromosomes are both efficient and relatively simple. Integrons also provide the machinery that is required to ensure that the new genes are expressed (Munita and Arias 2016).

5.4.4 Misuse and Overuse

Antibiotic resistance is boosted by antibiotic usage practices by humans. Epidemiological studies reveal that misuse, overuse, and irrational prescription lead to the spread of resistant bacterial strains. The increase in virulence can be due to antibiotic-induced gene expression, while an increase in HGT and mutagenesis help spread antibiotic resistance. Improper knowledge and misconceptions regarding antibiotic use can lead to an increase in antibiotic-resistant bacterial strains. Furthermore, overuse, irrational prescriptions, and the use outside clinics/hospitals are major causes of antibiotic resistance. Self-medication is indeed one of the common causes of antibiotic medication (Ventola 2015; Bu-Khamsin et al. 2021).

5.4.5 Unprocessed Disposal

The inappropriate disposal of used, stored, or expired antibiotics is a major concern. People are unaware of the proper disposal of antibiotics and thus end up flushing or throwing them away. Such improper disposal practices lead to the entry of antibiotics into landfills, drains, and water supplies. Due to the costs that are incurred with the correct disposal of those medications, it has been reported that expired and unused antibiotics are also being thrown away in garbage cans, sewers, and water fills. Though antibiotics are disposed of at negligible concentration, bacteria can acquire antibiotic resistance through modifying targets or HGT. Herein, repositioning of the host genome occurs leading to the acquisition and subsequent transmission of resistance. Therefore, this careless release of antibiotics into the environment results in a reduction of their efficiency and increases the prevalence of antibiotic resistance, as previously harmless microbes mutate into dangerous pathogens that are resistant to treatment (Anwar 2020).

5.5 MECHANISMS FOR ANTIBIOTIC RESISTANCE

Bacteria develop smarter machinery to avoid the effects of antibiotics. They develop a "thief door" for protection against antibiotics and gain resistance against them.

5.5.1 Enzymatic Degradation of Antibiotics

Bacterial enzymes belonging to different classes target antibiotics by various mechanisms. The first mechanism includes enzymes targeting the antibiotic molecule. Bacteria can also develop antibiotic resistance by altering the antibiotic molecule structure. Bacterial enzymes such as hydrolases, transferases, and redox enzymes may be responsible for altering the structure of antibiotics which aid in the gain of resistance to antibiotics. Antibiotics comprise hydrolytically susceptible chemical bonds such as esters and amides that are responsible for biological activities. The enzymes produced by the bacteria target the exposed bonds, thus decreasing antibiotic activity. The amidase enzyme breaks the β-lactam ring of penicillin and cephalosporin antibiotics. Bacterial enzymes can also modify the cellular targets of antibiotic molecules (Egorov et al. 2018).

5.5.2 Alteration of Antimicrobial Target

Alteration of antibiotic targets prevents the binding of the drug, thus developing resistance. An excellent example of this kind of target is peptidoglycan. Bacterial peptidoglycan plays a crucial role in bacterial survival and hence is targeted by antibiotics. Because the target sites are vital for the cellular functions of organisms, it is impossible for microorganisms to circumvent the action of antimicrobials by completely displacing the target sites. The target may undergo mutational changes that render it less susceptible to inhibition, while the cell functions normally. The mutations in RNA polymerase and DNA gyrase are responsible for the attainment of resistance. To develop resistance to an antibiotic, it may be necessary in some cases to make additional modifications to the cell in order to compensate for the changed characteristics of the target site (Lambert 2005).

5.5.3 Alteration in Membrane Permeability

Gram-negative bacteria have a distinctive outer membrane that is made up of phospholipids, lipoproteins, lipopolysaccharides, and β-barrel porins. The outer membrane blocks the movement of potentially harmful substances like bile acid and antibiotics, thus serving as a secondary barrier. The outer membrane of Gram-negative bacteria is impermeable to antibiotics with a molecular weight of more than 1400 Da. Hence, novel antibiotics that target Gram-negative pathogens are hampered by one of the most significant outer membrane barriers.

Also, porin-based antibiotic resistance mechanisms are observed in clinical isolates having a Gram-negative nature. Bacteria have developed a number of resistance mechanisms in order to reduce their uptake of antibiotics. Some of these mechanisms include (i) porin expression downregulation; (ii) mutations in the region of eyelet and channel interior; and (iii) alternative porins expression (OmpF and OmpC exchange) (Choi and Lee 2019, Prajapati et al. 2021).

5.5.4 Genetic Mutation

Bacteria can change through genetic mutations and become resistant to antibiotic treatment if they are subjected to antibiotics at doses that are lower than those required to kill all of the bacterial cells in a population. A natural variety of mutations that change resistance is present throughout the entire process of mutating bacteria. The insertion of a plasmid into the chromosome that codes for mutant genes contributes to the enhancement of these genetic alterations leading to antibiotic resistance. The chromosomal mutations give rise to antibiotic resistance. Antibiotic resistance to numerous structurally and mechanistically diverse classes is facilitated by mutations in the ribosomal machinery genes. This improves the survivability of the heat shock protein as well as other membrane stresses. These mutant genes influence ribosome assembly, which results in widespread modifications at the transcriptomic and proteomic levels. These ribosomal alterations speed up the development of specific targeted resistance. Furthermore, the presence of wild-type growth factors leads to suppressor mutations (Navved et al. 2019).

5.5.5 Modify or Bypass the Antibiotic's Target

Alterations to the geometrical and chemical composition of the target within bacterial cells are two additional ways that the antibiotic-target interaction can be stopped. In any other case, the bacteria have the ability to attach a side chain of a chemical group to the drug structure, thereby protecting themselves from the antibiotic. There are times when bacteria are able to harvest a diverse variant of geometry that they require. For instance, bacteria that are resistant to vancomycin make a cell wall that is associated with bacteria that are predisposed. This particular type of cell wall is immune to the ability of antibiotics to penetrate it (Navved et al. 2019).

5.6 ANTIBIOTIC RESISTANCE AND ITS DETRIMENTAL EFFECT ON THE ENVIRONMENT

The aquatic ecosystem is dynamic in the context of habitats, plant species, microorganisms, and water conditions. The bioactive compounds of antibiotics have a major impact on non-target organisms. The antibiotic-sensitive group of microorganisms is affected negatively by the effect of antibiotics. The bacteriostatic or bactericidal effect due to antibiotics changes the microbial dynamics of an aquatic system. It has been discovered that antibiotic pollution in aquatic environments decreases the diversity of all microbes, including the taxa involved in carbon cycling and primary productivity.

The antibiotics can bring structural changes in the microbial population, hamper ecological function in the environment, and also change biogeochemical cycles. Hence, this results in disturbed biological processes in the aquatic system. The low concentration of antibiotics can also have a negative effect on microbial processes such as nitrogen transformation, sulfate reduction, methanogenesis, nutrient cycle, and organic matter degradation. The selective pressure imposed by antibiotic-resistant bacterial populations brings changes in phenotypic and genotypic characteristics, gene expression, and physiological traits. It has been found that antibiotic pollution in aquatic ecosystems increases the prevalence of harmful cyanobacterial species, which in turn causes freshwater eutrophication and poses a threat to human health. Antibiotic pollution has not only affected microbial diversity of the aquatic system but also the higher organisms inhabiting the aquatic system.

It has been evidenced that sub-inhibitory concentrations of macrolides can cause malformations in embryos of zebrafish, such as yolk sac enlargement and a deflated swim bladder. These concentrations can also affect the frequency of embryonic spontaneous movement.

Furthermore, antibiotics can have an effect on higher organisms by causing disruptions in the microbial populations that are normally associated with animal hosts. The microbiomes of animals perform a variety of important functions, ranging from efficient nutrient metabolism to the enhancement of bone formation for the host. Thus, dysbiosis, i.e. disruption of a host-microbiome, can lead to health consequences, such as allergies, developmental defects, increased susceptibility to pathogens, and metabolic diseases. Antibiotic side effects are quite likely to affect the fish microbiome, which is extremely vulnerable to environmental change. The antibiotics used in aquaculture at low concentrations as a preventative measure can lead to a reduction in the variety of microorganisms found in the fish guts and an increase in mortality rates. Fish exposed to WWTP effluent waters contaminated with a variety of chemicals, including high concentrations of antibiotics, were found to have a higher mortality rate (Kraemer 2019, Araújo et al. 2021).

5.7 TECHNIQUES FOLLOWED FOR REMOVING ANTIBIOTICS

The water bodies function as reservoirs for drinking water. Hence, the quality of water must be of threshold quality. The water containing antibiotic residues may compromise the quality of water. Hence, various techniques are practiced to remove antibiotics from water. Figure 5.2 describes various conventional, advanced, and bioremediation techniques practiced for removing antibiotics from water bodies.

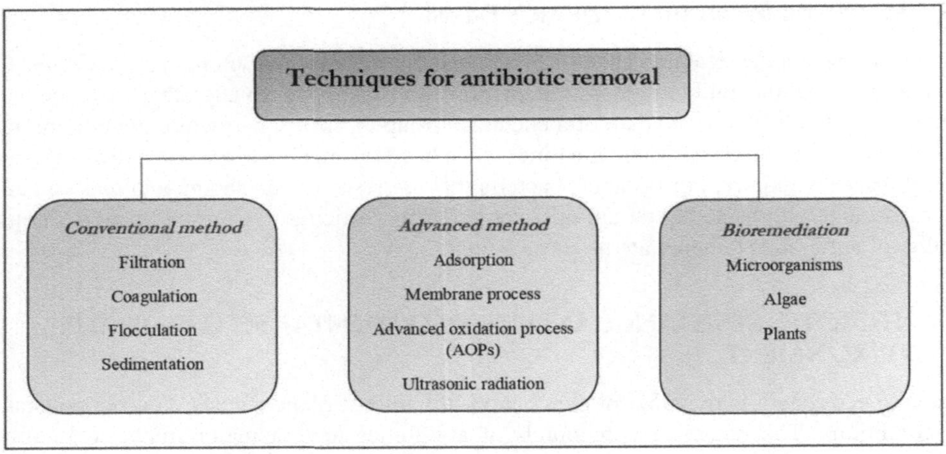

FIGURE 5.2 Techniques for antibiotic removal.

5.7.1 Conventional Technologies

5.7.1.1 Filtration, Coagulation, Flocculation, and Sedimentation

Conventional techniques are mostly used in the removal of antibiotics from WWTP. The process of filtration is carried out by forcing the wastewater to flow through a granular medium comprising solids such as sand, coal, diatomaceous earth, or granular activated carbon. The suspended matter can be particularly separated by the filtration process. Interstitial straining is another method followed for removing particles from the medium. However, smaller particles are retained. Smaller particles must be carried to the surface and held there by attachment mechanisms such as a chemical connection, electrostatic attraction, or adsorption in order to be removed. The contaminant is not degraded by this process; rather, it is concentrated in the solid phase, which results in the undesirable outcome of new waste.

Chemicals are utilized in the processes of sedimentation, flocculation, and coagulation in order to expedite the settling out of solids, the precipitation of pollutants, and the formation of colloids. Lime, alum, iron salts, and polymers are the chemicals that are utilized the most. These methods involve a subsequent treatment to be performed in order to get rid of the pollutants that are left over in the effluents. During the course of the past decade, a variety of tried-and-true approaches to antibiotic removal from environmental matrices have been put through their paces. The low efficiency of these procedures, which are occasionally ineffective, has led to the development of innovative alternatives (Homem and Santos 2011).

5.7.2 Advanced Technologies

5.7.2.1 Adsorption

Due to its simple design, flexibility, suitability for batch and continuous processes, low capital cost, and potential for regeneration and reuse, adsorption is widely recognized as an efficient, effective, and affordable method for the removal of various pollutants in the water system. The percentage removal rate of pollutants is comparatively high in the adsorption process. Sawdust is one example of a low-cost adsorbent that has been used in the past and is included on the list of available adsorbents. To date, numerous types of adsorbents have been developed. Biochars, cellulose, date palm leaflets, and corn bracts are all components that can be used. Other types of adsorbents, such as agricultural waste, lotus stalk, and rice husk, are currently being utilized as cost-effective substitutes for more pricey adsorbents. In order to increase the absorption efficiencies the pretreatment of chemical activation or thermal activation is given to the surface of the absorbent. The process of absorption can be divided into four phases:

Phase I: Bulk solute transportation
Phase II: Adsorbate film diffusion
Phase III: Adsorbate pore diffusion
Phase IV: Adsorbate and adsorbent interactions

Absorption of antibiotics by electrostatic interaction, hydrogen bonding, pore filling, and hydrophobic effect can be observed in the case of carbon-based adsorbents (Dutta and Mala 2020).

5.7.2.2 Membrane Processes: Reverse Osmosis, Nanofiltration, and Ultrafiltration

The membrane process is used increasingly for the separation process. The membrane separation can be classified mainly based on its application and the type of membrane used. The broad classification includes (i) mechanism of separation (ii) type of membrane used (iii) membrane geometric shape (iv) structure of the membrane. The separation processes mainly include reverse osmosis, nanofiltration, and ultrafiltration.

5.7.2.2.1 Reverse Osmosis

Semi-permeable spiral-wound membranes are utilized in the process of reverse osmosis in order to separate and then remove dissolved solids. The feed water is pushed through the semi-permeable membrane under a significant amount of pressure. Once it reaches the other side of the membrane, it travels through the membrane's minute pores to become the permeate water, which is then delivered as purified water. The reject stream is where the impurities in the water are collected, and the water is subsequently flushed down the drain. These membranes are semi-permeable, which means that they are able to prevent the passage of salt ions while still permitting the passage of water molecules. Reverse osmosis membranes can be made from a wide range of hollow-fibre materials, such as cellulose acetate, polyamides, and other types of polymers. The ability of the membrane to treat the water depends on the chemical composition of the feed water as well as the operational parameters of the plant (Garud et al. 2011).

5.7.2.2.2 Nanofiltration

A pressured stream of water is forced across a semi-permeable membrane. The membrane pore size allows only water molecules to pass by pressure gradient driving force thus separating them from residual matter. The separation of antibiotics mainly takes place on the basis of the ionic charge and size of the antibiotic molecule. The membranes can differ depending on the structure and pore size. The different membranes available are liquid membranes, non-porous, ceramic, dense, asymmetric, and charged. A tighter membrane has proven to have 99% removal efficiency as compared to a looser membrane. The nanofiltration process has the advantages of higher processing output, low loss, convenient maintenance, small footprint, compact structure, simple operation, low energy consumption, recycling, automation, and good stability (Akhil et al. 2021, Shi et al. 2022).

5.7.2.2.3 Ultrafiltration

The separation in the ultrafiltration process takes place on the basis of load size and particle size. The membrane pore with a size of 10–1000 angstrom can bring separation of molecules ranging in size from 300–500000 Dalton. The driving force with a pressure of 0.1~0.5 MPa is applied across the membrane in order to physically separate residues of different sizes. The advantages of ultrafiltration include higher efficiency, convenient operation process, small footprint, and dried format sludge production by air float that is not easily decayable. Also, the oxygen content of the water is increased due to ultrafiltration (Derakhshan et al. 2016, Shi et al. 2022).

5.7.2.2.4 *Microfiltration*

The microfiltration membrane has a pore size of 1–10 microns. The residues larger than the pore size are retained thus helping in separation. The microbial cells are unable to bypass the membrane pores. The microfiltration process has the advantages of high filtration accuracy, low capital investment, no secondary pollution, and greater reliability (Derakhshan et al. 2016, Shi et al. 2022).

5.7.2.3 Advanced Oxidation Process (AOP)

AOP techniques use oxidizing agents such as hydroxyl or sulfate radicals for purifying water resources. The oxidizing agents are known to degrade recalcitrant organic residues giving rise to less toxic compounds thus aiding in water purification. the different types of AOP includes Hydroxyl Radical-Based AOPs

5.7.2.3.1 *Hydroxyl Radical-Based AOPs*

The most preferred oxidizing agent in water remediation is the hydroxyl radical because of its reactive properties. It has an oxidation potential of between 2.8 V and 1.95 V at pH 0 and 14 respectively. The nonselective nature of hydroxyl radicals helps react with different pollutants at a rate constant of 10^8–$10^{1°}M^{-1}$ s^{-1}. Hydroxyl radicals degrade organic pollutants by the addition of radicals, the abstraction of hydrogen, electron transfer, and a radical combination that gives rise to producing carbon-centred radicals. This carbon-centred radical when reacted with oxygen gives rise to organic peroxyl radicals. This in turn leads to the degradation or mineralization of organic pollutants. The hydroxyl radicals are short-lived and can be produced in situ by a combination of oxidizing agents such as H_2O_2 and O_3, catalysts such as Fe^{2+}, and irradiation such as ultraviolet light or ultrasound.

5.7.2.3.2 *Ozone-Based AOPs*

The oxidant with the 2.07 V oxidizing potential is ozone. The selective process of direct ozonation has a rate constant of $1.0\times10^{°}$–10^3 M^{-1} s^{-1} [8]. Direct ozonation does not react with neutral forms of organic compounds but rather prefers reacting with ionized or dissociated forms. Hence, hydroxyl radical is produced from an indirect mechanism of indiscriminate oxidation. The hydroxyl radical can be produced by the following processes:

- In the peroxone system, ozone decomposition and hydroxyl production are enhanced by hydroperoxide (HO_2^-) produced from H_2O_2.
- In ozone-ultraviolet irradiation, ozone photolysis gives rise to H_2O_2 as an additional oxidant.

5.7.2.3.3 *UV-based AOPs*

UV-based AOPs produce hydroxyl radicals with the help of oxidants or catalysts. The catalyst commonly used is RO-type semiconductor titanium dioxide. Titanium oxide produces a positively charged hole in the valence band and a negatively charged electron in the conduction band with oxidative and reductive capacities respectively. The surface reaction of OH^-, H_2O, and O_2^- on titanium oxide gives rise to a hydroxyl group. Also, an H_2O_2 molecule can be broken using UV radiation to yield two hydroxyl radicals. Photolysis of an H_2O molecule at a wavelength of less than 242 nm can also give rise to hydroxyl radicals.

5.7.2.3.4 *Fenton-related AOPs*

The Fenton-related AOPs use metals to produce hydroxyl radicals by activating H_2O_2 in water. Iron metal is the most frequently used Fenton-related AOP where H_2O_2 reacts with Fe^{2+} ion producing a strong reactive species of hydroxyl radical. Hydroxyl radicals can be produced by the electron transfer chain as well as the scavenging of a Fenton reagent. The optimal molar ratio of Fe^{2+} ion: H_2O_2 has to be determined experimentally to reduce undesired scavenging. The iron sludge produced throughout the process restricts its application in wastewater treatment due to its high complexity and operational cost.

5.7.2.3.5 Sulfate radical-based AOPs

$S_2O_8^{2-}$ has a standard oxidation potential of 2.01 V and hence is a strong oxidant. The sulfate radical can be activated by means of heat, transitional metals, ultraviolet irradiation, or elevated pH. The activated sulfate radical has a standard oxidation potential of 2.6 V that can initiate sulfate radical-based AOPs. The persulfate can be activated by elevated pH or a temperature range of 35–130 °C. The metal activation method using the same molar persulfate concentration yields 50% sulfate radical and hence is theoretically inefficient. The metals used for activation are ferrous (Fe (II)), ferric (Fe (III)) ions, Cu (I), and Ag (I). Sulfate radicals remove electrons from organic pollutants and transfer them to organic radical cations. In alkaline conditions, sulfate radicals generate hydroxyl radicals (Deng and Zhao 2015).

5.7.2.3.5.1 Ultrasonic radiation Ultrasonic waves are mechanical waves that have a frequency of 20 HZ-20 KHZ. They have different properties of deflection, reflection, permeability, and diffraction. The frequency of waves, the density of waves, the type of pollutants, the nature of pollutants, the temperature of the water, and the water matrix.

5.8 BIOREMEDIATION

5.8.1 Antibiotic Bioremediation by Microorganisms

The physical and chemical method for antibiotic removal releases harmful byproducts at the end of the process. These harmful byproducts are more recalcitrant and toxic. Also, the antibiotics are stable at different temperatures and pH making them difficult to degrade. Thus, microorganisms could be used as a bioremediation tool for antibiotic removal. Microorganisms isolated from different niches have been proven to degrade antibiotic molecules. Antibiotic-degrading bacteria follow different metabolic pathways to metabolize the antibiotic molecule. They make use of pollutants as a sole source of carbon for energy. Also, the presence of antibiotic-resistant genes sometimes helps in the degradation of antibiotic molecules. The commonly used paracetamol drug is removed by a physical and chemical method such as an electrochemical process, UV oxidation, ozonation, titanium photocatalysis, etc. The bioremediation of paracetamol has been reported using *Penicillium* and bacterial species like *Pseudomonas species*, *Burkholderia species*, and *Stenotrophomonas species.* The isolation and identification of the aforementioned consortium has been successfully done for bioremediation. The bioremediation of antibiotics is affected by various biotic and abiotic factors. The biotic factors include species of culture, type of culture monoculture/consortium, and inoculum size. The abiotic factors include carbon sources, and environmental factors such as temperature, pH, salinity, agitation, time, and antibiotic type and concentration. Focusing on factors such as antibiotic-resistant genes frequency, biodegradation potential, redox mechanisms, and carbon and nutritional requirements would help isolate environment-friendly microbial consortium for the bioremediation of antibiotics (Singh and Saluja 2021, Warsito 2022).

5.8.2 Antibiotic Bioremediation by Algae

Algal cultures were initially used to remove nutrients and carbon sources from the aquatic system. Recently, algal-based bioremediation techniques are gaining interest among researchers. Algal cultures are the preferred agents for bioremediation as they remove contaminants and are ecological indicators.

They have short growth cycles, initiate stress response mechanisms, and are highly sensitive to aquatic pollutants. Also, they do not need additional energy for antibiotic removal from the water system. The other benefits offered by algal cultures include CO_2 fixation, solar energy-driven activity, and lower environmental impact. Algae is a potential source of raw materials for biofuel manufacturing. Antibiotic exposure causes the algal cultures to initiate a stress response. Various mechanisms followed by algal cultures for antibiotic removal are bio-adsorption, bioaccumulation, biodegradation, volatilization, photodegradation, and hydrolysis. Algae following the bio-adsorption mechanism have an adsorptive capacity wherein the pollutants are adsorbed by cell walls or extracellular polymeric substances discharged into the surrounding aquatic environment by algal cells. The uptake of antibiotics based on different biological, physical, and chemical mechanisms is known as bioaccumulation. The biodegradation mechanism by algal cultures includes metabolizing complex antibiotic compounds into simple compounds. A recent advancement has been reported wherein algae-activated sludge was used to efficiently remove cephalosporin at a removal efficiency of 97.91% (Li et al. 2022).

5.8.3 Antibiotic Bioremediation by Plants

Constructed wetland consists of plants, soil, and microorganisms that are artificially built to treat the wastewater. The water is purified by filtration, adsorption, coprecipitation, and ion exchange. Further plants help in the adsorption of antibiotic residues, and microorganisms degrade them. Constructed wetlands are economical, have broad-spectrum decontamination capabilities, and are efficient, easily manageable, and maintainable. The plant used for remediation purposes could be natural or genetically modified. The phytoremediation of antibiotics is a promising approach due to its cost-effective and environmentally friendly advantages. The study was carried out using *Pteris vittata (L.)* plant species for tetracycline removal from water. *Pteris vittata (L.)* was able to remove half the concentration of antibiotics on day 1 of the treatment. Day 5 of treatment showed no traces of tetracycline in the water. The plant followed the uptake/degradation in the fronds mechanism for phytoremediation of antibiotics (Jagtap 2020).

5.9 CONCLUSION

Antibiotics have a wide range of applications in the field of medicine. But the overuse and misuse of antibiotics have raised serious concerns about the same. The improper disposal and misuse of antibiotics have caused a detrimental effect on the environment. Also, antibiotic resistance amongst bacteria is a growing concern that needs to be addressed. Various physical and chemical methods are available for the remediation of antibiotics from freshwater. But these methods have disadvantages, such as the use of high energy, chemicals, and the production of sludge containing toxic secondary byproducts. Hence, there is a need to combat the antibiotic resistance era using greener tools such as bacteria, fungi, algae, and plants. Bioremediation using these greener agents would be an eco-friendly approach towards a sustainable environment.

REFERENCES

Akhil, D., Lakshmi, D., Senthil Kumar, P., Vo, D. V. N., & Kartik, A. (2021). Occurrence and removal of antibiotics from industrial wastewater. In *Environmental Chemistry Letters* (Vol. 19, Issue 2). Springer International Publishing. https://doi.org/10.1007/s10311-020-01152-0

Anwar, M., Iqbal, Q., & Saleem, F. (2020). Improper disposal of unused antibiotics: an often overlooked driver of antimicrobial resistance. *Expert Review of Anti-Infective Therapy*, *18*(8), 697–699. https://doi.org/10.1080/14787210.2020.1754797

Araújo, L. C. A. de, Silva, S. M. da, Sá, R. A. de Q. C. de, Lima, A. V. A., Barbosa, A. V., Silva, J. dos S., … Oliveira, M. B. M. de. (2021). Effects of antibiotics on impacted aquatic environment microorganisms. *Emerging Contaminants*. doi: 10.5772/intechopen.93910

Arsène, M. M. J., Davares, A. K. L., Viktorovna, P. I., Andreevna, S. L., Sarra, S., Khelifi, I., & Sergueïevna, D. M. (2022). The public health issue of antibiotic residues in food and feed: Causes, consequences, and potential solutions. In *Veterinary World* (Vol. 15, Issue 3, pp. 662–671). https://doi.org/10.14202/vetworld.2022.662-671

Bhutta, Z. A. (2006). Current concepts in the diagnosis and treatment of typhoid fever. *British Medical Journal*, *333*(7558), 78–82. https://doi.org/10.1136/bmj.333.7558.78

Bu-Khamsin, A. Y., Fabella, E., Al Abdullah, A. H., Albather, M. R., Al-Hassan, Y. T., Alramadan, H. A., Al Moweshy, A. A., Al Abdullah, A. J., & Al Hassan, H. I. (2021). Association between awareness on antibiotic resistance and antibiotic misuse among Saudi university students. *The Open Public Health Journal*, *14*(1), 545–554. https://doi.org/10.2174/1874944502114010545

Chausali, N., Jyoti Saxena, J., Prasad, R. (2022). Recent trends in nanotechnology applications of bio-based packaging. *Journal of Agriculture and Food Research*. https://doi.org/10.1016/j.jafr.2021.100257

Choi, U., & Lee, C. R. (2019). Distinct roles of outer membrane Porins in antibiotic resistance and membrane integrity in Escherichia coli. *Frontiers in Microbiology*, *10*(953), 1–9. https://doi.org/10.3389/fmicb.2019.00953

Danner, M. C., Robertson, A., Behrends, V., & Reiss, J. (2019). Antibiotic pollution in surface fresh waters: Occurrence and effects. *Science of the Total Environment*, *664*, 793–804. https://doi.org/10.1016/j.scitotenv.2019.01.406

Deng, Y., & Zhao, R. (2015). Advanced Oxidation Processes (AOPs) in wastewater treatment. *Current Pollution Reports*, *1*, 167–176. https://doi.org/10.1007/s40726-015-0015-z

Derakhshan, Z., Mokhtari, M., Babaei, F., Ahmadi, R. M., Ehrampoush, M. H., & Faramarzian, M. (2016). Removal methods of antibiotic compounds from aqueous environments- A Review. *Journal of Environmental Health and Sustainable Development*, *1*(1), 51–74.

Dutta, J., & Mala, A. A. (2020). Removal of antibiotic from the water environment by the adsorption technologies: A review. *Water Science and Technology*, *82*(3), 401–426. https://doi.org/10.2166/wst.2020.335

Egorov, A. M., Ulyashova, M. M., & Rubtsova, M. Y. (2018). Bacterial enzymes and antibiotic resistance. *Acta Naturae*, *10*(4), 33–48.

Gao, Y., Shang, Q., Li, W., Guo, W., Stojadinovic, A., & Mannion, C. (2020). Antibiotics for cancer treatment: A double-edged sword. *Journal of Cancer*, *11*(17), 5135–5149. https://doi.org/10.7150/jca.47470

Garud, R. M., Kore, S. V, Kore, V. S., & Kulkarni, G. S. (2011). A short review on process and applications of reverse Osmosis. *Universal Journal of Environmental Research and Technology*, *1*(3), 233–238.

Guardabassi, L., Schwarz, S., & Lloyd, D. H. (2004). Pet animals as reservoirs of antimicrobial-resistant bacteria. *Journal of Antimicrobial Chemotherapy*, *54*(2), 321–332. https://doi.org/10.1093/jac/dkh332

Hasan, C. M., Dutta, D., & Nguyen, A. N. T. (2021). Revisiting antibiotic resistance: Mechanistic foundations to evolutionary outlook. *Antibiotics*, *11*(1), 1–23. https://doi.org/10.3390/antibiotics11010040

Homem, V., & Santos, L. (2011). Degradation and removal methods of antibiotics from aqueous matrices – A review. *Journal of Environmental Management*, *92*(10), 2304–2347. https://doi.org/10.1016/j.jenvman.2011.05.023

Jagtap, Umesh. (2020). Bioremediation Strategies for Removing Antibiotics from the Environment. *Antibiotics and Antimicrobial Resistance Genes, Emerging Contaminants and Associated Treatment Technologies* (pp 319–337). 10.1007/978-3-030-40422-2_15

Karlo, J., Prasad, R., Singh, S. P. (2023). Biophotonics in food technology: Quo vadis? *Journal of Agriculture and Food Research*. https://doi.org/10.1016/j.jafr.2022.100482

Kırmusaoğlu, S., Gareayaghi, N., & S. Kocazeybek, B. (2019). Introductory chapter: The action mechanisms of antibiotics and antibiotic resistance. *Antimicrobials, Antibiotic Resistance, Antibiofilm Strategies and Activity Methods*, 1–9. https://doi.org/10.5772/intechopen.85211

Koch, N., Islam, N. F., Sonowal, S., Prasad, R., Sarma, H. (2021). Environmental antibiotics and resistance genes as emerging contaminants: Methods of detection and bioremediation. *Current Research in Microbial Sciences* 2, 100027. https://doi.org/10.1016/j.crmicr.2021.100027

Kraemer, S. A., Ramachandran, A., & Perron, G. G. (2019). Antibiotic pollution in the environment: From microbial ecology to public policy. *Microorganisms*, *7*(6), 1–24. https://doi.org/10.3390/microorganisms7060180

Lambert, P. A. (2005). Bacterial resistance to antibiotics: Modified target sites. *Advanced Drug Delivery Reviews*, *57*(10), 1471–1485. https://doi.org/10.1016/j.addr.2005.04.003

Li, S., Show, P. L., Ngo, H. H., & Ho, S. H. (2022). Algae-mediated antibiotic wastewater treatment: A critical review. *Environmental Science and Ecotechnology*, *9*(100145). https://doi.org/10.1016/j.ese.2022.100145

Manyi-Loh, C., Mamphweli, S., Meyer, E., & Okoh, A. (2018). Antibiotic use in agriculture and its consequential resistance in environmental sources: Potential public health implications. In *Molecules* (Vol. 23, Issue 4). https://doi.org/10.3390/molecules23040795

McManus, P. S., Stockwell, V. O., Sundin, G. W., & Jones, A. L. (2002). Antibiotic use in plant agriculture. *Annual Review of Phytopathology*, *40*(18), 443–465. https://doi.org/10.1146/annurev.phyto.40.120301.093927

Munita, J., & Arias, C. (2016). Mechanisms of antibiotic resistance. *Microbiol Spectr*, *4*(2), 1–37. https://doi.org/10.52403/ijrr.20221254

Navved, M., Chaudhry, Z., Bukhari, A., Meer, B., & Hajra, A. (2019). Antibiotics resistance mechanism. *Antibiotics and Antimicrobial Resistance Genes in the Environment*, *1*, 292–309. https://doi.org/10.1016/B978-0-12-818882-8.00019-X

Osińska, A., Korzeniewska, E., Harnisz, M., Felis, E., Bajkacz, S., Jachimowicz, P., Niestępski, S., & Konopka, I. (2020). Small-scale wastewater treatment plants as a source of the dissemination of antibiotic resistance genes in the aquatic environment. *Journal of Hazardous Materials*, *381*(August 2019), 121221. https://doi.org/10.1016/j.jhazmat.2019.121221

Prajapati, J. D., Kleinekathöfer, U., & Winterhalter, M. (2021). How to enter a bacterium: Bacterial Porins and the Permeation of antibiotics. *Chemical Reviews*, *121*(9), 5158–5192. https://doi.org/10.1021/acs.chemrev.0c01213

Prasad, R., Siddhardha, B., Dyavaiah, M. (2020). *Nanostructures for Antimicrobial and Antibiofilm Applications*. Springer International Publishing (ISBN 978-3-030-40336-2). www.springer.com/gp/book/9783030403362

Shi, Y., Wang, X., Feng, C., & Songtao, Z. (2022). Technologies for the removal of antibiotics in the environment: A review. *International Journal of Electrochemical Science*, *17*. https://doi.org/10.20964/2022.07.74

Silva, C. P., Louros, V., Silva, V., Otero, M., & Lima, D. L. D. (2021). Antibiotics in aquaculture wastewater: Is it feasible to use a photodegradation-based treatment for their removal? *Toxics*, *9*(8). https://doi.org/10.3390/toxics9080194

Singh, A., & Saluja, S. (2021). Microbial Degradation of Antibiotics from Effluents. In *Recent Advances in Microbial Degradation* (pp. 389–404). Springer. https://doi.org/10.1007/978-981-16-0518-5_15

Sivaraman, S., & Parady, V. (2018). Antibiotic use in food animals: India overview. In ReAct Asia-Pacific (pp. 1–31). www.reactgroup.org/wp-content/uploads/2018/11/Antibiotic_Use_in_Food_Animals_India_LIGHT_2018_web.pdforg

Sosa-Hernández JE, Rodas-Zuluaga LI, López-Pacheco IY, Melchor-Martínez EM, Aghalari Z, Limón DS, Iqbal HMN, Parra-Saldívar R. (2021). Sources of antibiotics pollutants in the aquatic environment under SARS-CoV-2 pandemic situation. *Case Studies in Chemical and Environmental Engineering*. Dec;*4*,100127. doi: 10.1016/j.cscee.2021.100127

Ventola Lee. (2015). The antibiotic resistance crisis. In *Comprehensive Biochemistry* (Vol. 40, Issue 4, pp. 277–283). https://doi.org/10.1016/B978-1-4831-9711-1.50022-3

Warsito, M. F. (2022). A review on potential microbes for bioremediation of antibiotics contamination in the environment. *IOP Conference Series: Earth and Environmental Science*, *1017*(012026), 1–8. https://doi.org/10.1088/1755-1315/1017/1/012026

6 Production of Antimicrobial Substances from Endophytic Microorganisms

Potential Applications as Biocontrol Agents in Agriculture

Pranay Jain and Ram Kumar Pundir

6.1 INTRODUCTION

Endophytes are referred to as the microorganisms such as fungi, bacteria and actinomycetes that colonize the inter- and/or intracellular location of plants such as the seed, leaf, stem and roots and they exhibit a symbiotic relationship with the plants (Pimentel et al. 2011, Singh and Dubey 2015, Yadav 2018). All or part of the life cycle of endophytes occurs within the hosts, without causing any apparent symptoms of diseases. They are ubiquitous in nature and exhibit complex interactions with their hosts involving mutualism, antagonism and rarely parasitism (Nair and Padmavathy 2014). Endophytes are responsible for improving plant growth by secretion of phytohormones and help in overall nutritional growth using bidirectional nutritional transfer and protecting plants from attack of plant pathogens (Andreozzi et al. 2019, Shen et al. 2019). These endophytes have been found in many plants such as rice, wheat, tomato, cowpea, maize, strawberry, chickpea, mustard, sugarcane, chili, citrus, soybean, cotton, pearl millet and sunflower (Verma et al. 2017, Yadav et al. 2018). Endophytes have gained attention as of great biotechnological and industrial relevance owing to their ability to produce secondary metabolites, by serving as biocontrol agents, antimicrobial agents, antitumour agents and immunosuppressents. They also produce antiviral compounds, antioxidants, antidiabetic compounds, antibiotics and agriculturally relevant insecticidal compounds (Gouda et al. 2016, Yadav 2018). Owing to the development of microbicide resistance in plant pathogens and ecological and public health considerations, the number of available chemical microbicides is being reduced. In addition, the industry has to deal with the challenges in the form of increasing costs for the development and commercialization of novel disease-control products. For these reasons, there is an increasing need for the development of alternative disease management products, which can offer environmentally friendly and economically feasible control of important crop plant diseases. Along with the selection of resistant plants and other management practices, biological control by beneficial microorganisms is increasingly considered an important element in integrated disease management strategies and there has been an intensified search for endophytes in recent years (Latz et al. 2018). This review emphasizes that the antimicrobial substances produced by endophytic fungi, bacteria and actinomycetes have potential as biocontrol agents in agriculture.

DOI: 10.1201/9781003268895-6

6.2 ANTIMICROBIAL SUBSTANCES PRODUCTION BY ENDOPHYTES

Most endophytic fungi, bacteria and actinomycetes produce secondary metabolites and some of them exhibit antimicrobial properties which help in inhibiting the growth of phytopathogenic microorganisms (Gunatilaka 2006). Different bioactive compounds have been identified globally from endophytes for their ability to inhibit many phytopathogens (Suryanarayan 2013, Daguerre et al. 2016). Secondary metabolites with antimicrobial properties from endophytes include flavonoids, peptides, quinones, alkaloids, phenols, steroids, terpenoids and polyketides (Mousa and Raizada 2013, Lugtenberg et al. 2016).

6.3 ANTIMICROBIALS FROM ENDOPHYTIC FUNGI

Endophytic fungi represent a rich source of specialized metabolites and other compounds that may be inhibiting competing microorganisms (Thines et al. 2004). Researchers all over the world have identified and developed fungal metabolites or the producing fungi towards commercial application in plant disease control and a range of different compounds have been investigated for their antimicrobial activity against phytopathogens (Schulz et al. 2002, Thines et al. 2004, Suryanarayanan 2013, Daguerre et al. 2016).

Kusari et al. (2012) suggested that the presence of different microbial species in the same plant can trigger metabolite production by both endophyte and/or the host to limit the growth of potentially pathogenic organisms. It has been found that the host and endophyte share parts of a specific pathway and contribute partially to the metabolite production or one partner induces metabolism or metabolizes products of the other as evident in the case of the production of the anticancer drug taxol in the Pacific yew tree *Taxus brevifolia* where several fungal endophytes were associated with the production (Ludwig-Müller 2015). Laur et al. (2018) suggested that often the effects of metabolites are assessed under *in vitro* conditions which gives an indication of the potential of an organism to produce substances that could inhibit the pathogen, it does not give an indication pathogen inhibition will take place under *in vivo* conditions. Mathivanan et al. (2008) are of the opinion that while direct application of *in vitro*-produced metabolites may control disease in the plant by direct toxicity or by inducing resistance the endophyte may not actually produce these compounds in planta. This is due to major differences in nutrient levels, environmental conditions as well as interactions with other organisms in the *in vivo* situation, with complex interactions between biocontrol agent, plant, inherent microbiome and pathogen. Kusari et al. (2012) revealed that different kinds of artificial media or modification of other factors can give rise to the production of different types of metabolites.

Lugtenberg et al. (2016) and Gabriel et al. (2018) suggested that for demonstrating whether an antimicrobial compound produced *in vivo* is involved in biocontrol induced by an endophyte, ideally it should be confirmed that the compound is in contact with the pathogen. This may be difficult to determine since the endophyte is localized inside the plant and the amount of metabolite produced may be very low. Furthermore, metabolites can be translocated in the plant from the endophyte to the site of pathogen infection, whereas volatile organic compounds produced by a biocontrol agent could diffuse to the site of the pathogen infection.

Silva et al. (2006) isolated an endophytic fungi *Phomopsis cassiae* from *Cassia spectabilis*, which was able to synthesize sesquiterpenes that were most active against *Cladosporium cladsporioides* and *C. sphaerospermum*.

Hellwig et al. (2002) isolated altersetin, a novel alkaloid from the endophytic fungi, *Alternaria* spp. which was found to have strong potential in inhibiting the proliferation of many pathogenic Gram-positive bacteria.

Muscodor albus, an endophytic fungus from tropical trees has been known to produce many volatile organic compounds, including aciphyllene, 2-butanone and 2-methyl furan which have exhibited antibiotic properties (Atmosukarto et al. 2005). Liu et al. (2001) reported that fungal endophytes isolated under *in vitro* conditions from *Artemisia annua* could suppress the growth of most phytopathogenic organisms by the production of antifungal compounds such as n-butanol and ethyl acetate. Tian et al. (2017) reported the role of the antifungal protein produced by *Epichloë festucae* in controlling *Sclerotinia homoeocarpa* in *Festuca rubra*.

Mejia et al. (2008) isolated endophytic fungi *Colletotrichum gloeosporioides*, *Clonostachys rosea* and *Botryosphaeria ribis* from *Theobroma cacao* and screened them under *in vitro* conditions for antimicrobial activity against pathogenic fungi *Moniliophthora roreri*, *Phytophthora palmivora* and *Moniliophthora perniciosa*, respectively. Nonetheless, 13%, 21%, and 0% of tested morphospecies showed clear antibiosis against *M. roreri*, *P. palmivora*, and *M. perniciosa*, respectively. Field trials assessing the effects of three endophytic fungi on pod loss due to *M. roreri* and *Phytophthora* spp. were conducted at four farms in Panama. Although the overall incidence of black pod rot was very low during the tests, treatment with *C. gloeosporioides* significantly decreased pod loss due to that disease. Treatment with *C. rosea* reduced the incidence of cacao pods with sporulating lesions of *M. roreri* by 10%. The observed reduction in pod loss due to *Phytophthora* spp., and sporulation by *M. roreri*, suggested the potential of fungal endophytes as biological control agents.

Monteiro et al. (2017) identified endophytic fungi from *Coffea arabica* (coffee plant) that could produce volatile organic compounds and evaluated the effect of the VOCs produced by endophytic fungi on the growth of *Rhizoctonia solani*, *Fusarium oxysporum*, *Phoma* sp., *Botrytis cinerea*, *Fusarium solani*, *Fusarium verticillioides*, *Cercospora coffeicola* and *Pestalotia longisetula*, and selected endophytic fungi with potential for biological control of *Aspergillus ochraceus* inoculated in coffee beans and *F. verticillioides* inoculated in corn seeds. An isolate of *Muscodor albus* was used as a selection tool for volatile organic compounds producing fungi. Among the 400 endophytic fungi isolates, 11 were able to grow in the presence of volatile organic compounds produced by *M. albus*. These fungi were identified as *Muscodor* spp. and *Simplicillium* sp. according to searches in the UNITE database using DNA sequences of internal transcribed spacer (ITS). The volatile organic compounds produced by endophytic fungi inhibited the growth of the phytopathogenic fungi with different efficacies, compared to the control. The volatile organic compounds produced by *Muscodor coffeanum* showed a fungicidal effect against *A. ochraceus* on coffee beans. Six endophytic fungi completely inhibited the growth of *F. verticillioides* inoculated in corn seeds. This study demonstrated that the volatile-compound-producing endophytic fungi, isolated from *Coffea arabica* could be used as promising sources of bioactive compounds.

Endophytic fungi from *Nyctanthes arbor-tristis* were isolated and evaluated for their antimicrobial activity by Gond et al. (2012). A total of 19 endophytic fungi were isolated from 400 segments of healthy leaf and stem tissues of *N. arbor-tristis*. The leaf endophytes *Colletotrichum dematium* inhibited 55.87% of the radial growth of the phytopathogen *Curvularia lunata*.

Ambrose and Belanger (2012) isolated a purified form of *Efe-AfpA* antifungal protein mined from an apoplastic fluid of endophyte-inoculated red fescue which exhibited anti-parasitic activity against *Sclerotinia homoeocarpa*. The same result was also observed in the recombinant product of *the Efe-AfpA*-expressed gene found in *Pichia pastoris*. In a transcriptome study to detect the percentage protein (*Efe-AfpA*) produced from the endophytic relationship between *Epichloë festucae* and *Festuca rubra* sp., a 6%—*Efe-AfpA* was produced from the fungal transcriptome.

Singh et al. (2021) evaluated the antimicrobial activity and phytochemical analysis of endophytic fungus *Fusarium proliferatum*, recovered from the root tissues of a folk medicinal plant *Cissus quadrangularis L.*, which was identified by microscopic and molecular methods. The fungus showed broad-spectrum antibacterial and antifungal activities against a variety of important plant pathogens such as *Rhizoctonia solani* and *Fusarium oxysporum*. The minimum inhibitory concentration of the crude extract ranged from 0.2–2.5 mg/ml against fungal pathogens.

Santos et al. (2019) studied the biotechnological potential of endophytic fungi, isolated from leaves of *Sapindus saponaria*, to control phytopathogenic fungi. Molecular taxonomy was performed by sequencing the ITS1-5.8S-ITS2 ribosomal DNA region, identifying the genera *Phomopsis*, *Sordariomycetes*, *Diaporthe* and *Colletotrichum*. *In vitro* antagonism against phytopathogens showed better results against *Fusarium solani* and provided inhibition indices between 41.8 % and 67.5 %. The endophytic fungus, *Diaporthe citri* presented the highest antagonism index against the pathogen. Against *Glomerella* sp. and *Moniliophthora perniciosa*, inhibition rates ranged between 18.7 % and 57.4 % and between 38.3 % and 64.8 %, respectively.

Ortiz-Ojeda et al. (2020) isolated endophytic microorganisms from *Acmella ciliate*, a native herb of South America and evaluated their ability to inhibit pathogenic fungi. A total of 56 fungi and 39 endophytic bacteria, were isolated from the leaves. The endophyte isolates were then tested in antagonism assays against the phytopathogens *Fusarium decemcellulare* and *Colletotrichum gloeosporioides*. The best antagonism index values against *C. gloeosporioides* were obtained with the endophytic fungus (40%). The highest antagonism index values were obtained against *F. decemcellulare* with the fungal genera *Curvularia*, *Colletotrichum*, *Plectosphaerella* and *Sordariomycetes*.

6.4 ANTIMICROBIALS FROM ENDOPHYTIC BACTERIA

Endophytic bacteria associated with medicinal plants possess unique strategies that enhance the growth and survival of host plants, many of which are mediated by distinctive secondary metabolites. These bacteria and their secondary metabolites are important subjects for both basic and applied research aimed at sustainable agriculture.

In a study carried out by Mohammad et al. (2018), 114 endophytic bacterial strains isolated from the wild ethnomedicinal plant *Glycyrrhiza uralensis* (licorice) were screened for their *in vitro* antimicrobial activities against common fungal pathogens of tomato (*Fusarium oxysporum* f. sp., *Fulvia fulva*, *Alternaria solani*), cotton (*Fusarium oxysporum* f. sp. Vesinfectum, *Verticillium dahliae*), pomegranite (*Ceratocystis fimbriata*), *Cymbidinium* (*Colletotrichum gloeosporioides*), and Tsao-ko (*Pestalotiopsis microspora* and *Fusarium graminearum*). Several *Bacillus* strains, particularly *Bacillus atrophaeus* and *Bacillus mojavensis*, had a broad spectrum of antifungal and antibacterial activity. A total of 16 strains, selected based on broad antimicrobial activity, were shown to contain at least one putative secondary metabolite-encoding gene (i.e., polyketide synthase or non-ribosomal peptide synthetase) and/or one lytic enzyme (i.e., protease, cellulase, lipase, chitinase), which may be important mediators of antagonistic activity against pathogens. Five strains, representing *Bacillus atrophaeus* and *Bacillus mojavensis*, were selected for plant growth chamber experiments based on strong *in vitro* antifungal activities. All five strains significantly reduced disease severity in *Arabidopsis thaliana* plants challenged with *V. dahlia* infection. Gas-chromatography/mass-spectrometry analysis of cell-free extracts of *Bacillus atrophaeus* strain XEGI50 showed that at least 13 compounds were produced only during co-cultivation with *V. dahlia*, including putative compounds known to have antimicrobial activity, such as 1,2-benzenedicarboxylic acid, bis (2-methylpropyl) ester; 9,12-octadecadienoic acid (Z,Z)-, methyl ester; 9-octadecenoic acid, methyl ester, (E)-; and decanedioic acid, bis(2-ethylhexyl) ester. The study provided new insights into the antimicrobial activities of natural endophytes, particularly *B. atrophaeus*, and suggested that this species could be a promising candidate as a biocontrol agent to confer resistance to *Verticillium* wilt disease and other phytopathogens in cotton and other crops.

Bacterial endophytes from two medicinal plant species abundantly available in North Eastern states of India, Chirata (*Swertia chirata*) and Datura (*Datura stramonium*), had been isolated and studied for their antibacterial and antifungal activities by Devi et al. (2021). Six bacterial endophytes were isolated from Chirata and five from Datura. All these bacterial isolates were initially isolated by observing their colony morphology, and then nine morphologically distinct bacterial isolates were identified at the molecular level by their 16S rRNA sequencing. Among the identified isolates, five were found to belong to genus *Staphylococcus*, two *Bacillus*, and one each belonged to *Rhodococcus* and *Cupriavidus*. An antibacterial study was carried out against two plant pathogenic bacteria, viz., *Xanthomonas oryzae* and *Ralstonia solanacearum*. An antifungal study was carried out against two plant pathogenic fungi, viz., *Fusarium oxysporum* and *F. solani*. Out of six bacterial isolates from Chirata, one showed antibacterial as well as antifungal properties, and another one showed antifungal properties only. Likewise, out of five bacterial isolates from Datura, one was found to have both antibacterial and antifungal properties, and another one had only antifungal properties.

Lopes et al. (2018) identified and characterized endophytic bacteria isolated from the roots, stems and leaves of soyabean and evaluated them for their capacity to inhibit fungal and bacterial plant pathogens. Thirteen isolates showed *in vitro* antagonism against *Sclerotinia sclerotiorum*, *Phomopsis sojae* and *Rhizoctonia solani*. *Bacillus* sp. and *Burkholderia* sp. were the most effective isolates in controlling bacterial and fungal pathogens *in vitro*. Extracts and precipitates from culture supernatants of isolates showed different patterns of inhibitory activity on the growth of fungal and bacterial pathogens. *Bacillus* sp. and *Burkholderia* sp. were the most effective isolates in controlling fungal pathogens *in vitro*, and the activity is mainly due to peptides. However, most of the studied bacteria showed the presence of antimicrobial compounds in the culture supernatant, either peptides, bacteriocins or secondary metabolites. These results could be significant to develop tools for the biological control of soybean diseases. The work brought the identification of microorganisms such as *Bacillus* sp. and *Burkholderia* sp. that have the potential to protect crops in order to enhance a sustainable management system of crops.

Meliah et al. (2021) conducted a study to evaluate the antagonistic activity of endophytic bacteria recovered from sweet sorghum against *Sclerotium rolfsii*, *Fusarium solani*, *Fusarium oxysporum*, *Colletotrichum gloeosporioides in vitro* and evaluated the mechanisms of these fungal inhibitions. 78 endophytic bacteria were selected from the stem and root of sweet sorghum plants. They were tested for antagonist activity by direct confrontation method. Antifungal compound production was examined to determine their mechanisms in inhibiting fungal pathogens. Antifungal compound production was evaluated by detecting the presence of non-ribosomal peptides and polyketide genes. Selected bacteria were identified using molecular analysis based on the 16S rRNA gene. 14 out of the 78 tested isolates showed antagonistic activity and two were able to inhibit all four tested fungal strains. Endophytic bacteria identified as *Burkholderia* were revealed to have potential as a biocontrol agent.

Camellia oleifera, an oil-producing plant, is widely grown in warm, subtropical, hilly regions in China. However, *C. oleifera* is strongly negatively affected by *C. oleifera* anthracnose, which is caused by *Colletetrichum fructicola*. To find a suitable biocontrol agent for *C. oleifera* anthracnose, 41 endophytes were isolated from the stems, leaves, and roots of *C. oleifera* by Xu et al. (2020). Bacterial cultures were identified based on analyses of 16S rDNA sequences; most strains belonged to the genus *Bacillus*. The antagonistic effects of these strains on *C. fructicola* were tested *in vitro*. In total, 16 strains inhibited *C. fructicola* growth, with *B. subtilis* being the most efficient. This bacterial strain demonstrated antagonistic activity against *C. siamense*, *C. asianum*, *Fusarium proliferatum*, *Agaricodochium camellia* and *Pseudomonas syringae*.

In a study carried out by Liu et al. (2020), 37 culturable endophytic bacteria strains were obtained from potato tubers based on the culture-dependent method. Results indicated that nine strains showed antimicrobial activity against at least one pathogen by antimicrobial activity screening and 23 strains showed inhibitory activity on solanine in potato tubers. Among them, *Bacillus megaterium* showed not only better antimicrobial activity against both the two indicator pathogens but also the best inhibitory activity on solanine, which was proved to be a potential biocontrol bacterium. Meanwhile, the relationship between the distribution of the endophytic bacterial community and the content of solanine in potato tubers was studied by Illumina-based analysis, indicating that the distribution of the endophytic bacterial community was influenced by the content of the solanine. The results showed a new insight into the relationship between the plant's secondary metabolites and endophytic bacteria in potato tubers and provided potential new technical support for the biological control of potato storage.

In a study carried out by Panigrahi and Rath (2021), *Enterobacter cloacae* isolated from surface-sterilized parts of *Ocimum sanctum* showed potent antimicrobial activities against two fungal phytopathogens, *Pythium debaryanum* and *Rhizoctonia solani*. The isolate *E. cloacae* inhibited the growth of *Rhizoctonia solani* and *Pythium debaryanum*, while studied through dual-culture and well plate assay methods. Higher inhibition % was reported against *Rhizoctonia solani* (73.75 ± 2.5%) than *Pythium debaryanum* (60.24 ± 2.08%). Secondary metabolites like alkaloids, phenols, polychlorinated biphenyls, aromatic hydrocarbons, esters, etc. were identified in the extract through Gas Chromatography coupled Mass Spectrometry (GC-MS), which could be responsible for antimicrobial activities as observed during the investigation.

In a study carried out by Paul et al. (2013), endophytic bacteria that exhibited antagonism against phytopathogens were isolated from healthy tissues of leaves, stems and roots of chili pepper plants (*Capsicum annum* L.) in 2010–2011. Antifungal activities of all collected isolates were tested against plant pathogens by dual-culture method. Pathogenic fungi used in this study were *Alternaria panax*, *Botrytis cinerea*, *Colletrotichum acutatum*, *Fusarium oxysporum* and *Phytophthora capsici*. A diverse range of other bacterial taxa were isolated and identified- *Actinobacter*, *Arthrobacter*, *Enterobacter*, *Escherichia*, *Kitasatospora*, *Pandoraea*, *Pantoea*, *Rhizobium*, *Ralstonia*, *Paenibacillus*, and *Serratia*. Dual culture antifungal activity indicated that 22 bacterial isolates (12%) inhibited at least one pathogenic fungus tested. *Bacillus tequilensis*, *Burkholderia cepacia*, *Pseudomonas aeruginosa* showed antifungal activity against all tested fungi. Crude extracts of selected isolates showed antifungal activity against *Botrytis cinerea* and among others, *Burkholderia cepacia* performed the strongest antifungal activity (inhibition zone >55 mm) by paper disk method.

A study was carried out by Xiaojie et al. (2019) to assess the inhibitory activities of endophytic bacteria from *Eucommia ulmoides* against *Fusarium graminearum*. The endophytic bacteria from *E. ulmoides* were isolated and purified, and the antimicrobial activities of endophytic bacteria against *F. graminearum* were examined using a confront antibiotic culture experiment, and disease-control pot experiment. The results indicated 104 endophytic bacteria were isolated from the leaves, stems and roots of *E. ulmoides*, and *Bacillus* sp. had strong inhibition against *F. graminearum*, Inhibitory rate of *Bacillus* sp. on *F. graminearum* was 69.04%. Bacterial suspension and secondary metabolite of *Bacillus* sp. had a strong teratogenic effect on the conidia of *F. graminearum* conidia.

Research was conducted by Pitiwittayakul et al. (2021) to explore endophytic bacteria with antifungal activity against *Fusarium moniliforme*. In this study, 52 isolates of endophytic bacteria associated with the roots and stems of sugarcane from Nakhon Ratchasima province, Thailand were obtained. *In vitro* antagonistic activity test showed that 14 out of 52 isolates had antagonistic activity against the fungal pathogen *F. moniliforme*. These antagonistic endophytic bacteria were identified as belonging to six different species as follows: *Nguyenibacter vanlangensis*, *Acidomonas methanolica*, *Asaia bogorensis*, *Tanticharoenia aidae*, *Burkholderia gladioli* and *Bacillus altitudinis* based on phenotypic characteristics, along with phylogenetic analysis of their 16S rRNA gene sequences. Seven isolates effectively inhibited *F. moniliforme* mycelial growth by up to 40%. The volatile compounds of six isolates reduced the growth of *F. moniliforme* by over 23%. These results demonstrated that the tested endophytic bacteria could be successfully utilized as a source of biocontrol agent to manage diseases caused by *F. moniliforme*.

6.5 ANTIMICROBIALS FROM ENDOPHYTIC ACTINOMYCETES

Endophytic actinomycetes have attracted microbiologists and agrochemists as promising producers of novel antimicrobials and growth promoters to develop new medicines and agrochemicals. Although the fruits of basic research concerning these organisms have not sufficiently been realized, it is important to review the present knowledge of the beneficial interactions of endophytic actinomycetes and their hosts for their potential use in crop management to help solve a possible worldwide biological control of crops from various phytopathogens in the near future.

In a study carried out by Quecine et al. (2008), the chitinase production by endophytic actinomycetes was carried out, and the potential of chitinase for control of phytopathogenic fungi was investigated. Endophytic Streptomyces were grown on a minimum medium supplemented with chitin, and chitinase production was quantified. The strains were screened for any activity towards phytopathogenic fungi and oomycetes by a dual-culture *in vitro* assay. The correlation between chitinase production and pathogen inhibition was calculated and further confirmed *on Colletotrichum sublineolum* cell walls by scanning electron microscopy.

Singh and Gaur (2016) isolated sixty-eight chitinolytic endophytic actinomycetes from various medicinal plants and evaluated them for their chitinase activity. Among these isolates, 12 were screened for their plant-growth-promoting abilities and antagonistic potential against *Scleoritium rolfsii*. Further, these isolates were validated *in vivo* for their ability to protect chickpeas against *S. rolfsii* infestation under greenhouse conditions. The isolates significantly (P < 005) increased the biomass (12–20 fold) and reduced plant mortality (42–75%) of chickpeas. On the basis of 16S rDNA profiling, the selected antagonistic strains were identified as *Streptomyces diastaticus*, *Streptomyces fradiae*, *Streptomyces olivochromogenes*, *Streptomyces collinus*, *Streptomyces ossamyceticus* and *Streptomyces griseus*. The isolated species showed potential for controlling collar rot disease on chickpeas and could be useful in integrated control against diverse soil-borne plant pathogens.

Zhuang et al. (2020) isolated an endophytic actinomycete with antagonistic activity against *Ralstonia solanocearum*, the causative agent of bacterial wilt of tomato, from a moss, *Physcomitrium sphaericum* (Ludw)(Fuernr). The biocontrol experiment demonstrated that co-inoculation by the isolate and the pathogen gave the greatest biocontrol efficiency of 63.6%. The *Streptomyces* strain had morphological characteristics and chemotaxonomic properties identical to *Streptomyces azureus* and *Streptomyces anandii*. The strain was proposed to be the novel species as *S. physcomitrii* sp. *nov*. Tan et al. (2006) assessed the biocontrol potential of endophytic actinomycetes against bacterial wilt of tomato caused by *Ralstonia solanacearum* and tested different isolates for their potential for production of siderophores.

Thirty-one endophytic streptomycete and non-streptomycete actinobacteria were isolated from healthy date palm root tissues by Alblooshi et al. (2022). *In vitro* screening revealed that the antifungal action of *Streptomyces polychromogenes* was associated with the production of cell-wall degrading enzymes, whereas with diffusible antifungal metabolites in *Streptomyces coeruleoprunus*, albeit their production of volatile antifungal compounds. The two antagonists were recovered from root tissues until 12 weeks after inoculation, and efficiently colonized root cortex and xylem vessels, indicating that the date palm roots are a suitable habitat for these endophytic isolates. At the end of the greenhouse experiments, the development of sudden decline syndrome (SDS) was markedly suppressed by 53% with the application of *S. polychromogenes* and 86% with *S. coeruleoprunus*, confirming their potential in disease management. Results showed that the estimated disease severity indices in diseased seedlings were significantly ($p < 0.05$) reduced from 4.75 (scale of 5) to 2.25 or 0.67 by either *S. polychromogenes* or *S. coeruleoprunus*, respectively. In addition, conidial numbers of the pathogen significantly ($p < 0.05$) dropped by 38% and 76% with *S. polychromogenes* and *S. coeruleoprunus*, respectively, compared to infected seedlings with *F. solani* (control). Thus, the suppression of disease symptoms was superior in seedlings pre-inoculated with *S. coeruleoprunus*, indicating that the diffusible antifungal metabolites were responsible for *Fusarium solani* retardation in these plants.

In a study carried out by Gangwar et al. (2011), 40 endophytic actinomycetes were isolated from roots, stems and leaves of three medicinal plants viz. *Aloe vera*, *Mentha arvensis* and *Ocimum sanctum*. The identification revealed that the majority of the isolates were *Streptomyces* spp. and the rest were identified as *Saccharopolyspora* spp., *Micromonospora* spp. and *Actinopolyspora* spp. The dual tests revealed that nine endophytic actinomycete isolates displayed a wide spectrum activity against nine fungal phytopathogens. Out of eight isolates, 90% inhibited the growth of at least one or more phytopathogenic fungi and *Saccharopolyspora* spp. exhibited antagonistic activity against *Aspergillus niger*, *Aspergillus flavus*, *Alternaria brassicicola*, *Botrytis cinerea*, *Penicillium digitatum*, *Fusarium oxysporum*, *Penicillium pinophilum*, *Phytophthora dresclea* and *Colletotrichum falcatum*.

Costa et al. (2013) studied the biodiversity of maize endophytic actinomycete, and an evaluation of their potential to control the phytopathogenic fungi. From several regions of São Paulo state, 40 strains were isolated from the healthy maize plants. The identification of these strains, based on morphological properties and fatty acid methyl ester (FAME) profile showed that most of them belonged to the *Streptomyces* genus. These isolates were first screened for the growth inhibition of phytopathogenic fungi and results showed that all the isolates were able to inhibit the development of at least one tested pathogen. Two selected endophytic actinobacterial isolates were then evaluated for the control of *Pythium aphanidermatum* in cucumber (*Cucumis sativa* L.) under greenhouse conditions.

A greenhouse investigation was carried out by ElTarabily et al. (2009) with three endophytic actinomycetes *Actinoplanes campanulatus*, *Micromonospora chalcea* and *Streptomyces spiralis* for their potential to protect cucumber from pathogen *Pythium aphanidermatum* causing damping-off, crown and root rot. These three endophytic isolates were screened on the basis of their ability to produce β-1,3, β-1,4 and β-1,6-glucanases to antagonize *P. aphanidermatum*. These endophytes were found to produce glucanase, especially in a consortial treatment which could be used in place of metalaxyl, a fungicide recommended for *Pythium* diseases in the area.

Dhanasekaran et al. (2005) mentioned several mechanisms of endophytic actinomycetes to protect the plant which involves the production of antifungal compounds, chitinolytic activities and competition for nutrients through siderophore production. Antibiosis is likely to be the important mechanism for the biocontrol activity of actinomycetes as most of the isolates that showed antagonism *in vitro* also showed its effect *in vivo* (Trejo-Estrads et al. 1998).

Verma et al. (2009) isolated endophytic actinomycetes from neem (*Azadirachta indica*). The most common genus found was *Streptomyces*. A few other genera isolated belonged to *Streptosporangium*, *Microbispora*, *Streptoverticillium*, *Sacchromonospora* and *Nocardia* were also isolated. These isolates had shown antagonistic activity against root pathogens *Pythium* and *Phytophthora* sp. and could be used as biocontrol agents against these fungal pathogens.

Rice endophytes were examined for their biocontrol potential by Tian et al. (2004). The biocontrol potential of endophytic fungi and actinomycetes was assessed. In dual culture with pathogens, 41.2% of endophytic fungi and 50% of endophytic actinomycetes were found antagonistic to fungal pathogens. The major genera in endophytic actinomycetes were found to be *Streptomyces griseofuscus* and *S. hygroscopicus*. More diversity of endophytic actinomycetes was found in the roots of rice plants and in alkaline soil.

The antimicrobial activity of endophytic actinomycetes was also studied in *Rhododendron* (Shimizu et al. 2000) and found effective against Gram-positive bacteria, yeast and filamentous fungi. Cao et al. (2004) deciphered biocontrol activity of endophytic actinomycetes against Panama wilt of banana. A few of the strains, like *Streptomyces griseorubiginosus* were described as potential biocontrol agents against Panama wilt pathogen *Fusarium oxysporum* f. sp. *cubense*. Siderophore producing *Streptomyces* endophytes were suggested as a biological control agent for *Fusarium* wilt of bananas (Cao et al. 2004).

6.6 FUTURE PROSPECTS

Concise studies on a specific population of endophytes active in a host are required before bulk production can be carried out which often requires research with advanced technology. Studies are also needed as regards getting plant-specific inoculum doses of endophytes, this will help in reducing bulk production and also enhance productivity by reducing dependence on synthetic fungicides, pesticides, and fertilizers. The development of endophytes that could be sprayed just like most chemical pesticides will help in the acceptance of endophytes in integrated pest management. Future studies shall be required to take into account the development of genomic tools and metabolomics tools for further studies on how endophytes colonize the plant and plant-microbe interaction. There is still a need to study the compounds produced by endophytes and their activities in reducing diseases. This will help in developing efficient markers for some important and distinct biocontrol agents and assessing the effects of plant genotypes, innate microbe community, and most importantly the environment. This structured approach will also help in discovering new endophytes with important traits.

6.7 ACKNOWLEDGEMENT

The authors are grateful to the Hon'ble Vice-Chancellor of Kurukshetra University, Kurukshetra and Additional Director, University Institute of Biotechnology, Chandigarh University, Mohali, Punjab for encouragement and providing the necessary infrastructure for carrying out research on entophytes.

REFERENCES

Alblooshi, A. A., G. P. Purayil, E. E. Saeed et al. 2022. Biocontrol potential of endophytic actinobacteria against *Fusarium solani*, the causal agent of sudden decline syndrome on date palm in the UAE. *J. Fungi*. 8: 8. doi: 10.3390/jof8010008

Ambrose, K. V., F. C. Belanger 2012. SOLiD-SAGE of endophyte-infected red fescue reveals numerous effects on host transcriptome and an abundance of highly expressed fungal secreted proteins. *PLoS ONE* 7: e53214. doi:10.1371/journal.pone.0053214s

Andreozzi, A., P. Prieto, J. Mercado-Blanco et al. 2019. Efficient colonization of the endophytes *Herbaspirillum huttiense* RCA24 and *Enterobacter cloacae* RCA25 influences the physiological parameters of *Oryza sativa* L. cv. *Baldo Rice. Environ. Microbiol.* 21, 3489–504. doi: 10.1111/1462-2920.14688

Atmosukarto, I., U. Castillo, W. M. Hess, J. Sears, G. Strobel. 2005. Isolation and characterization of *Muscodor albus* I-41.3s, a volatile antibiotic producing fungus. *Plant Sci.* 169, 854–861. doi: 10.1016/j.plantsci.2005.06.002

Cao, L., Z. Qiu, J. You, H. Tan, S. Zhou. 2004. Isolation and characterization of endophytic *Streptomyces* strains from surface-sterilized tomato (*Lycopersicon esculentum*) roots. *Lett. Appl. Microbiol.* 39: 425–430.

Costa, F. G., T. D. Zucchi, I. S. de Melo. 2013. Biological control of phytopathogenic fungi by endophytic actinomycetes isolated from maize (*Zea mays* L.). *Braz. Arch. Biol. Technol.* 56(6): 948–955.

Daguerre, Y., V. Edel-Hermann, C. Steinberg. 2016. Fungal genes and metabolites associated with the biocontrol of soil-borne plant pathogenic fungi. In: *Fungal Metabolites*, Springer International Publishing, 1001. doi: 10.1007/978-3-319-19456-1_27-1

Devi, R., T. Nath, R. R. Boruah et al. 2021. Antimicrobial activity of bacterial endophytes from Chirata (*Swertia chirata* Wall.) and Datura (*Datura stramonium* L.). *Egypt J. Biol. Pest Control* 31: 69 (2021). doi: 10.1186/s41938-021-00410-9

Dhanasekaran, D., P. Sivamani, A. Pannerselvam, N. Thajuddin, G. Rajakumar, S. Selvamani. 2005. Biological control of tomato seedling damping off with *Streptomyces* sp. *Plant Path. J.* 4: 91–95.

El-Tarabily, K. A., A. H. Nassar, G. Hardy, K. Sivasithamparam. 2009. Plant growth promotion and biological control of Pythium aphanidermatum, a pathogen of cucumber, by endophytic actinomycetes. *J. Appl. Microbiol.* 107: 672–681.

Gabriel, K. T., D. Joseph Sexton, C. T. Cornelison. 2018. Biomimicry of volatile-based microbial control for managing emerging fungal pathogens. *J. Appl. Microbiol.* 124(5): 1024–1031.

Gangwar, M., S. Dogra, N. Sharma. 2011. Antagonistic bioactivity of endophytic actinomycetes isolated from medicinal plants. *J.Adv.Lab. Res. Biol.* 2(4): 154–157

Gond, S., A, Mishra, V. K. Sharma et al. 2012. Diversity and antimicrobial activity of endophytic fungi from *Nyctanthes arbor-tristis*, a well-known medicinal plant of India. *Mycosci.* 53(2): 113–121.

Gouda, S., G. Das, S. K. Sen, H.-S. Shin, J. K. Patra. 2016. Endophytes: A treasure house of bioactive compounds of medicinal importance. *Front. Microbiol.* 7:1538. doi: 10.3389/fmicb.2016.01538

Gunatilaka, A. L. 2006. Natural products from plant-associated microorganisms: distribution, structural diversity, bioactivity and implications of their occurrence. *J. Nat. Prod.* 69: 509–526.

Hellwig, V., T. Grothe, A. Mayer-Bartschmid et al. 2002. Altersetin, a new antibiotic from cultures of endophytic *Alternaria* spp. *J. Antibiot.* 55: 881–892. doi: 10.7164/antibiotics. 55.881

Kusari, S, C. Hertweck, M. Spiteller. 2012. Chemical ecology of endophytic fungi: Origins of secondary metabolites. *Chem Biol.* 19(7): 792–798.

Latz, M. A. C., B. Jensen, D. B. Collinge, H. J. L. Jergensen. 2018. Endophytic fungi as biocontrol agents: Elucidating mechanisms in disease suppression. *Plant Ecol. Divers.* 11(5–6): 555–567.

Laur, J, G. B. Ramakrishnan, C. Labbé, et al. 2018. Effectors involved in fungal–fungal interaction lead to a rare phenomenon of hyperbiotrophy in the tritrophic system biocontrol agent–powdery mildew–plant. *New Phytol.* 217(2): 713–725.

Liu, C. H., W. X. Zou, H. Lu, R. X. Tan 2001. Antifungal activity of *Artemisia annua* endophyte cultures against phytopathogenic fungi. *J. Biotechnol.* 88, 277–282. doi: 10.1016/S0168-1656(01)00285-1

Liu, J-M., S-S. Wang, X. Zheng et al. 2020. Antimicrobial activity against phytopathogens and inhibitory activity on solanine in potatoes of the endophytic bacteria isolated from potato tubers. *Front. Microbiol.* doi: 10.3389/fmicb.2020.570926

Lopes, K. B. A., V. Carpentieri-Pipolo, D. Fira et al. 2018. Screening of bacterial endophytes as potential biocontrol agents against soybean diseases. *J. Appl. Microbiol.* 125: 1466–1481

Ludwig-Müller, J. 2015. Plants and endophytes: Equal partners in secondary metabolite production?. *Biotechnol Lett.* 37 (7): 1325–1334.

Lugtenberg, B. J., J. R. Caradus, L. J. Johnson. 2016. Fungal endophytes for sustainable crop production. *FEMS Microbiol. Ecol.* 92: 1–17. doi: 10.1093/femsec/fiw194

Mathivanan, N, V. R. Prabavathy, V. R. Vijayanandraj. 2008. The effect of fungal secondary metabolites on bacterial and fungal pathogens. In: *Secondary Metabolites in Soil Ecology*. P. Karlovsky ed. Springer. 129–140.

Mejia, L. C., E. I. Rojas, Z. Maynard, et al. 2008. Endophytic fungi as biocontrol agents of *Theobroma cacao* pathogens. *Bio. Control.* doi:10.1016/j.biocontrol.2008.01.012

Meliah, S., T. R. Sulistiyani, P. Lisdiyanti et al. 2021. Antifungal activity of endophytic bacteria associated with sweet sorghum (*Sorghum bicolor*). *J. Math. Fund. Sci.* 53(1): 16–30.

Mohammad, O. A. A., Li Lil, J. B. Ma et al. 2018. Evaluation of the antimicrobial activity of endophytic bacterial populations from chinese traditional medicinal plant licorice and characterization of the bioactive secondary metabolites produced by *Bacillus atrophaeus* against *Verticillium dahlia.* Front. Microbiol. doi: 10.3389/fmicb.2018.00924

Monteiro, M. C. P., N. M. Alves, M. Queiroz et al. 2017. Antimicrobial activity of endophytic fungi from coffee plants. *Biosci. J.* 33(2): 381–389.

Mousa, W. K., M. N. Raizada. 2013. The diversity of anti-microbial secondary metabolites produced by fungal endophytes: An interdisciplinary perspective. *Front. Microbiol.* 4: 65. doi: 10.3389/fmicb.2013.00065

Nair, D. N., S. Padmavathy. 2014. Impact of endophytic microorganisms on plants, environment and humans. *Sci. World. J.* 250693. doi: 10.1155/2014/250693

Ortiz—Ojeda, C. P., S. L. de Andrade, R. E. L. Procopio. 2020. Antifungal activity of endophytic microorganisms isolated from *Acmella ciliata* (Asteraceae). *Gen. Mol. Res.* 19(2): gmr18570.

Panigrahi, S., C. C. Rath. 2021. *In vitro* characterization of antimicrobial activity of an endophytic bacterium *Enterobacter cloacae* (MG001451) isolated from *Ocimum sanctum. South Afr. J. Bot.* 143: 90–96.

Paul, N. C., S. H. Ji, J. X. Deng, S. H. Yu. 2013. Assemblages of endophytic bacteria in chilli pepper (*Capsicum annuum* L.) and their antifungal activity against phytopathogens *in vitro*. *Plant Omics J.* 6(6): 441–448.

Pimentel, M. R., G. Molina, A. P. Dionisio, M. R. Marostica, G. M. Pastore. 2011. Use of endophytes to obtain bioactive compounds and their application in biotransformation process. *Biotechnol. Res. Int.* 576286. doi: 10.4061/2011/576286

Pitiwittayakul, N., D. Wongsorn, S. Tanasupawat. 2021. Characterisation of plant growth-promoting endophytic bacteria from sugarcane and their antagonistic activity against *Fusarium moniliforme. Trop. Life Sci. Res.* 32(3): 97–118.

Quecine, M. C., W. L. Araujo1, J. Marcon, C. S. Gai1, J. L. Azevedo1, A. A. Pizzirani-Kleiner. 2008. Chitinolytic activity of endophytic Streptomyces and potential for biocontrol. *Lett. Appl. Microbiol.*47: 486–491.

Santos, C. M. dos, A. S. Ribeiro, A. Garcia et al. 2019. Enzymatic and antagonistic activity of endophytic fungi from Sapindus saponaria L. (Sapindaceae). *Acta Biol Columb.* doi: 10.15446/abc.v24n2.74717

Schulz, B., C. Boyle, S. Draeger, K. Krohn. 2002. Endophytic fungi: A source of novel biologically active secondary metabolites. *Mycol Res.* 106(9): 996–1004.

Shen, F.-T., J.-H. Yen, C.-S. Liao, W.-C. Chen, Y.-T. Chao 2019. Screening of rice endophytic biofertilizers with fungicide tolerance and plant growth promoting characteristics. *Sustainability* 11: 1133. doi: 10.3390/su11041133

Shimizu, M., Y. Nakagawa, Y. Sato et al. 2000. Studies on endophytic actinomycetes (I) *Streptomyces* sp. isolated from *Rhododendron* and its antifungal activity. *J. Gen. Plant Pathol.* 66(4): 360–366.

Silva, G. H., H. L. Teles, L. M. Zanardi, et al. 2006. Cadinane sesquiterpenoids of *Phomopsis cassiae*, an endophytic fungus associated with *Cassia spectabilis* (Leguminosae). *Phytochemistry* 67: 1964–1969. doi: 10.1016/j.phytochem.2006.06.004

Singh, A., J. Kumar, V. K. Sharma et al. 2021. Phytochemical analysis and antimicrobial activity of an endophytic *Fusarium proliferatum* (ACQR8) isolated from a folk medicinal plant *Cissus quadrangularis* L. *South Afric. J. Bot.* 140: 87–94.

Singh, R., A. K. Dubey. 2015. Endophytic actinomycetes as emerging source for therapeutic compounds. *Indo Global J Pharm Sci.* 5: 106–116.

Singh, S. P., R. Gaur. 2016. Evaluation of antagonistic and plant growth promoting activities of chitinolytic endophytic actinomycetes associated with medicinal plants against *Sclerotium rolfsii* in chickpea. *J. Appl. Microbiol.* 121(2): 506–518.

Suryanarayanan, T. S. 2013. Endophyte research: Going beyond isolation and metabolite documentation. *Fungal Ecol.* 6: 561–568.

Tan, H. M., L. X. Cao, Z. F. He, G. J. B. Lin, S. N. Zhou. 2006. Isolation of endophytic actinomycetes from different cultivars of tomato and their activities against *Ralstonia solanacearum in vitro*. *World J. Microbiol. Biotechnol.* 22(12): 1275–1280.

Thines, E., H. Anke, R. W. S. S. Weber. 2004. Fungal secondary metabolites as inhibitors of infection-related morphogenesis in phytopathogenic fungi. *Mycol. Res.* 108(1): 14–25.

Tian, B., C. Zhang, Y. Ye et al. 2017. Beneficial traits of bacterial endophytes belonging to the core communities of the tomato root microbiome. *Agric. Ecosyst. Environ.* 247: 149–156. doi: 10.1016/j.agee.2017.06.041

Tian, X. L., L. X. Cao, H. M. Tan, H. M. et al. 2004. Study on the communities of endophytic fungi and endophytic actinomycetes from rice and their antipathogenic activities *in vitro*. *World J. Microbiol. Biotechnol.* 20(3): 303–309.

Trejo-Estrada, S. R., I. R. Sepulveda, D. L. Crawford 1998. *In vitro* and *in vivo* antagonism of *Streptomyces violaceunsniger* YCED9 against fungal pathogen of turf grass. *World J. Microbiol. Biotechnol.* 21: 81–90.

Verma, P., A. N. Yadav, V. Kumar, D. P. Singh, A. K. Saxena. 2017. Beneficial plant-microbes interactions: Biodiversity of microbes from diverse extreme environments and its impact for crop improvement. In *Plant-Microbe Interactions in Agro-Ecological Perspectives*, eds D. P. Singh, H. B. Singh, and R. Prabha (Singapore: Springer), 543–580.

Verma, V. C., S. K. Gond, A. Kumar et al. 2009. Endophytic actinomycetes from *Azadirachta indica* A. Juss.: Isolation, diversity, and anti-microbial activity. *Microb. Ecol.* 57(4): 749–756.

Xiaojie, C., W. Qi, Z. Xinyue, D. Ting. 2019. Study on endophytic bacteria isolated from *Eucommia ulmoides* with antimicrobial activity against *Fusarium graminearum*. *Acta Agric.Zhejian.* 31(5): 766–776.

Xu, J-X, Z-Y. Li, X. Lv. et al. 2020. Isolation and characterization of Bacillus subtilis strain 1-L-29, an endophytic bacteria from *Camellia oleifera* with antimicrobial activity and efficient plant-root colonization. *PLoS ONE* 15(4): e0232096. doi: 10.1371/ journal.pone.0232096

Yadav, A. N. 2018. Biodiversity and biotechnological applications of hostspecific endophytic fungi for sustainable agriculture and allied sectors. *Acta Sci. Microbiol.* 1: 44.

Zhuang, X., C. Gao, C. Peng et al. 2020. Characterization of a novel endophytic actinomycete, *Streptomyces physcomitrii* sp. nov., and its biocontrol potential against *Ralstonia solanacearum* on tomato. *Microorganisms.* 8: 2025. doi:10.3390/microorganisms8122025

7 Antimicrobial Efficacy of Medicinal Plants

Present Status and Future Prospects

Aniket Singh[*] *and Neela Singh*
[*] Corresponding author: dr.aniketsingh89@gmail.com

7.1 INTRODUCTION

The World Health Organization (WHO) has declared that 80% of developing countries profit from the utilization of conventional medicines that are derived from medicinal plants (Palhares 2015). In the entire world, the expected number of plants is around 374,000, and among those, humans use 28,187 medicinal species (Christenhusz and Byng 2016; MNPS 2023). Twenty thousand species of medicinal plants and their medicinal potential have been documented by the WHO (Yadav and Agarwala 2011). There are more than 1340 plants with distinct antimicrobial efficacy, and over 30,000 antimicrobial phytochemicals have been isolated from plants (Tajkarimi et al., 2010).

Antimicrobial resistance emerges due to misuse and overuse of antibiotics, making a lot of currently available medicine ineffective (Davies and Devies 2010; Baym et al. 2016). WHO has declared this emerging trend the most serious problem facing global public health (WHO 2015). Consequently, there is a rising demand to develop novel antimicrobial agents to fight against resistance development. So, researchers are trying to isolate and identify novel bioactive phytochemicals from plants to combat against microbial resistance (Khameneh et al. 2016; Tortorella et al. 2018). Medicinal plants are sources of countless bioactive compounds; a few of them have been isolated and identified by researchers (Cowan 1999; Saranraj and Sivasakthi 2014; Singh et al. 2019; Singh and Chandra 2022). Bioactive compounds can act singly or synergistically together or in mixture with antibiotics to increase antimicrobial efficacy against a broad range of microbes (Bazzaz et al. 2016; Bazzaz et al. 2018).

7.2 ANTIMICROBIAL EFFICACY OF MEDICINAL PLANT EXTRACTS

Different biological activities such as antioxidant, antimicrobial, and anti-inflammatory of medicinal plants extracts have been documented by different researchers (Ghosh et al. 2008; Dib et al. 2021; Obaid et al. 2022). Medicinal plant-derived phytochemicals inhibit the growth of bacteria, virus, protozoa and fungi by various mechanisms and are also helpful in the treatment of resistant microbial strains (Khameneh et al. 2016; Ruddaraju et al. 2020). Phytochemicals have enormous medicinal prospects as they have fewer side effects compared to synthetic agents and also lower chances of developing resistance (Kursa et al. 2022).

Bacteria may develop resistance to phytochemicals if only a single active fraction with a particular aim is involved, similar to an antibiotic (Reker et al. 2014). The effectiveness of phytochemicals to

DOI: 10.1201/9781003268895-7

hamper bacterial growth is also associated with the synergistic effect among the active fractions of the plant extracts. The synergistic action comes from various effects, such as the appearance of multi-target mechanisms capable of suppressing bacterial resistance mechanisms, and physico-chemical or pharmacokinetic effects resulting in improved solubility, resorption rate, reduced toxicity, and neutralizing the adverse effects (Wagner and Ulrich-Merzenich 2009).

Nanoparticles synthesized from different medicinal plant extracts along with different metal ions (copper, gold, silver, platinum, etc.) i.e., "green synthesis nanoparticles" showed effective biological activity as antibacterial, antiviral, anti-pathogenic, antifungal, etc. (Prasad., 2014; Ali et al. 2020). Green synthesis copper nanoparticles synthesized using *Ephedra alata* aqueous plant extract showed remarkable antibacterial activity against Gram-negative bacteria (*Staphylococcus aureus* and *Bacillus subtilis*) and antifungal activity against *Saccharomyces cerevisiae* and *Candida albicans* (Atri et al. 2023). Green synthesis silver nanoparticles synthesized from medicinal plant *Commelina forskaolii* extract exhibited strong antibacterial activity against *Pseudomonas aeruginosa* and *Enterococcus fecalis* and antifungal activity against *Aspergillus niger* and *Candida albicans* (Vijay et al. 2023). Green synthesis gold and platinum nanoparticles synthesized using medicinal plant extracts of *Catharanthus roseus* and *Taraxacum laevigatum*, respectively, showed effective antibacterial activity against Gram-positive bacteria *Bacillus subtilis* and Gram-negative bacteria *Pseudomonas aeruginosa* (Tahir et al. 2017; Jeevanantham et al. 2023).

Bioactive compounds of Cameroonian plants containing flavonoids, phenolics, alkaloids, steroids, and triterpenes showed effective antimicrobial activity (Victor 2010; Dzotam and Kuete 2017). *Azadirachta indica* and *Melia azedarach* contain phenolic and flavonoid compounds that exhibit strong antiviral activity against SARS-CoV-2, which is responsible for the global pandemic in early 2020. These two plants also showed significant antibacterial activity against Gram-positive (*Listeria monocytogenes*, *Enterococcus faecalis*, and *Staphylococcus aureus*) and Gram-negative bacteria (*Pseudomonas aeruginosa*, *Escherichia coli*, and *Salmonella enterica*) (Hemdan et al. 2023). Different bioactive compounds such as carotenoids, quercetin, spermidine, tocophero, and rutin derived from caper (*Capparis sp.)* are responsible for anti-inflammatory, antimicrobial, and antioxidative efficacy. *Capparis deciduas* seed extracts exhibited antifungal, antibacterial, and antileishmanial efficacy, possibly due to glucosinolate and quaternary ammonium (Tlili et al. 2011). *Salvia hispanica*, an herbaceous plant, contains enormous amounts of terpenoids and flavonoids that show effective antibacterial and antifungal activity (Motyka et al. 2023). The leaf extracts of *Verbena officinalis* and *Myrtus communis* exhibited effective antibacterial efficacy against *Salmonella typhi*, *Staphylococcus aureus*, and *Escherichia coli*. Tea tree (*Melaleuca alternifolia*) oil and the seed oil of carrot (*Daucus carota*) show potential antimicrobial activity against *Mycoplasma pneumoniae* and *Helicobacter pylori*, respectively (Wangchuk et al. 2011). Phytochemicals isolated from lemon balm (*Melissa officinalis*), tea tree (*Melaleuca alternifolia)*, and garlic (*Allium sativum*) are documented as broad-spectrum antimicrobial compounds (Joshi 2017). Hydromethanolic extracts of *Punica granatum,Cistus monspeliensis*, and *Berberis vulgaris* showed potential antibacterial efficiency against *Enterobacter cloacae*, *Staphylococcus aureus*, and *Enterococcus faecalis* (Bereksi et al. 2018). Also, methanol extracts of *Cinnamomum tamala*, *Ageratina adenophora*, *Oxalis corniculata*, and *Artemisia vulgaris* demonstrated antimicrobial efficiency against *Citrobacter koseri*, *Escherichia coli*, *Salmonella typhi*, *Staphylococcus aureus*, and *Klebsiella pneumoniae* (Manandhar, Shisir, Dahal 2019). Ethanol extract of aerial parts of *Helianthemum salicifolium* showed remarkable antibacterial activity against *Enterococcus faecalis*, *Pseudomonas aeruginosa*, *Acinetobacter baumannii*, *Staphylococcus aureus*, and *Escherichia coli*, as well as fungicidal activity against *Candida albicans*, *Candida krusei*, *Candida glabrata* (Mohammed et al. 2021). *Allium sativum* and *Caryophyllus aromaticus* extracts showed outstanding antibacterial efficacy against different phytopathogens (*Colletotrichum coccodes*, *Fusarium oxysporum*, *Fusarium sambucinum*, *Pectobacterium carotovorum*, etc.) of *Solanum tuberosum* (Steglinska et al. 2022).

Medicinal plant-derived resins revealed significant antimicrobial activity (Termentzi, Nikolas and Skaltsounis 2011). In the *Hypericum acmosepalum* plant, an endophytic fungus lives that contains a number of compounds such as emodin, hyperphorin, hyperenone A, and hyperenone B responsible for antibacterial potentiality against resistant *Staphylococcus aureus*, *Mycobacterium tuberculosis*, *Pseudomonas aeruginosa*, *Escherichia coli*, *Salmonella enterica*, *Klebsiella pneumoniae*, and also on the fungal strains *Candida albicans* and *Aspergillus niger* (Osman et al. 2012). The methanol solvent extract of *Hypericum olympicum* exhibited significant antimicrobial efficacy against *Salmonella enteritidis* and *Klebsiella pneumoniae* (Shiu et al. 2012). Medicinal aromatic plants, such as fennel, thyme, lavender, and peppermint-derived essential oils containing mixtures of volatile substances such as phenylpropanoids, monoterpenes, and sesquiterpenes, have been documented to be effective against Gram-positive and Gram-negative bacteria, viruses, and fungi (Reichling et al. 2009, Sienkiewicz et al. 2011). Phytochemical diaporthalasin isolated from the fungus *Diaporthaceae sp.* showed effective antibacterial efficiency against both methicillin-resistant *Staphylococcus aureus* (MRSA) and *Staphylococcus aureus* (Jin et al. 2016).

7.3 MECHANISMS OF ACTION OF ANTIMICROBIAL COMPOUNDS

Antimicrobial compounds work by stopping the vital cellular processes of bacteria and/or avoiding the usual method of antibacterial resistance. However, bacteria can generate resistance to numerous antimicrobial agents inherently by selective pressure or by getting equipment for resistance from adjacent microorganisms (Magiorakos et al. 2012).

7.3.1 Inhibition of Bacterial Protein Biosynthesis

Antibiotics such as tetracyclines, oxazolidinones, macrolides, and aminoglycosides kill bacteria by inhibiting protein synthesis. Amikacin attaches strongly to 16S ribosomal RNA (r RNA) and the S12 protein of the ribosomal 30S small subunit. As a result, the shape of the ribosome changes, making it unable to read messenger RNA (mRNA) codons accurately and ultimately inhibiting protein synthesis (Christopher 2003).

7.3.2 Inhibition of Replication and Transcription

DNA gyrase is an essential enzyme for the replication, transcription, and repair processes of bacteria. Antibiotics like nalidixic acid target the gyrase enzyme to inhibit the vital cellular procedures of bacteria. Ciprofloxacin works by stopping the functions of gyrase and topoisomerase IV and, as a result, stopping cell division (Anthony 1997).

7.3.3 Inhibition of the Cell-wall Biosynthesis

Antibiotics like vancomycin, penicillin, and cephalosporins bind to the cell wall peptidoglycan layer and inhibit different enzyme functions. These antibiotics inhibit the function of enzymes, transpeptidases and transglucosilases, that perform essential functions in bacterial cell wall formation, and as a result, bacteria fail to form cell walls properly (Schneider and Sahl 2010).

7.3.4 Destruction of the Bacterial Cell Wall

Polymyxins bind with lipopolysaccharides and cause structural modification of the bacterial outer membrane that results in osmotic imbalance and finally, the death of the bacteria. Antibiotics and polymyxin B alter the permeability of the bacterial outer membrane by binding the lipopolysaccharide layer, which destabilizes the outer membrane of bacteria (Tenover 2006).

7.4 MECHANISMS OF RESISTANCE TO ANTIBACTERIAL AGENTS

Bacteria develop resistance to antimicrobial agents by using various mechanisms, which are described in detail below.

7.4.1 Efflux Pump

Efflux pumps (EPs) work as an efflux or export system that can generate resistance to antibacterial drugs. Antibacterial drugs work after arriving at the target site of function at definite concentrations. Bacteria using EPs pump out the antibacterial agent before it has time to diffuse in their cells, and therefore, the intracellular concentration reduces its effective dose (Levy 1992). EPs are able to pass on both amphipathic and lipophilic molecules from the bacteria. They are also capable of pumping out a single type of compound and a range of structurally unlike antibacterial drugs, which are usually observed in multiple drug-resistant bacteria (Webber and Piddock 2003).

In bacteria, five main families of EPs have been documented: multidrug and toxic efflux (MATE), small multidrug resistance (SMR), resistance-nodulation division (RND), ATP binding cassette (ABC), and major facilitator superfamily (MFS) (Lomovskaya 2001). In both Gram-positive and Gram-negative bacteria, the SMR, MFS, MATE, and ABC families are mostly obtained, although the RND family is exclusively obtained in Gram-negative bacteria (Mahamoud et al. 2007).

Antibiotic resistance using this mechanism can be found in an extensive range of pathogenic Gram-positive and Gram-negative bacteria and fungi, including *Pseudomonas aeruginosa*, *Staphylococcus aureus*, *Candida albicans*, and *Acinetobacter baumannii* (Tegos et al. 2011).

7.4.2 Structural Modification of Porins

Porins make water-filled open channels that permit the passive transport of compounds across the lipid bilayer membrane. In this way, the influx of molecules into the cell is chiefly controlled by porins. Intracellular entrée of an antibiotic drug can be controlled by decreasing the antibiotic influx. Modification in the porins structure results in a change in membrane permeability, thus reducing antibiotic influx (Vila et al. 2007). Among many, this is one of the tactics of bacteria for antibacterial resistance that is often found in Gram-negative bacteria such as *Pseudomonas spp.* and *Acinetobacter spp.* (James and Winterhalter 2008).

7.4.3 Damaging the Antibacterial Agents

Another tactic of bacterial resistance is the chemical degradation of antibacterial agents or antibiotics. β-lactamase enzyme of bacteria by hydrolysis breaks the β-lactam ring of antibiotics (e.g., penicillins, carbapenems and cephalosporins) and as a result they become inactive (Olsen 2015).

7.4.4 Modification of Antibiotics

Antibiotic aminoglycosides are deactivated by the modification of functional groups at three different sites using three different types of modifying enzymes. This modified antibiotic has slowed a considerably lower affinity for RNA and is unable to bind to the ribosome and, as a result, is unable to stop bacterial protein synthesis (Shaw et al. 1993).

7.4.5 Altered Target

Bacteria can alter drug-binding sites so that the antibacterial drug is unable to react with it, and as a result, a significant reduction in antibacterial efficacy. This type of resistance is obtained in a broad range of resistant bacteria (Christopher 2000). Erythromycin resistance bacteria are attributed to the modification of a drug function site, which is the enzyme peptidyl transferase. Through the methylation of this enzyme, the process of translation that is mediated by antibiotics does not hamper but the affinity of the antibiotics reduces at the action site. This mechanism of resistance is documented in the drug-resistant bacteria *Staphylococcus aureus* (Bussiere et al. 1998).

7.5 MECHANISMS OF ACTION AND ANTIMICROBIAL EFFICACY OF MEDICINAL PLANT-DERIVED PHYTOCHEMICALS

From prehistoric times, humans have been using medicinal plants for the treatment of various diseases (Moloney 2016). Medicinal plants have massive prospects for the discovery of novel bioactive compounds which can combat against resistant microorganisms (Vaou et al. 2021). Phytochemicals of healing value are generally secondary metabolites of plants used for medicinal purposes. Plants' secondary metabolites are the outcome of their secondary metabolism (Verpoorte 1998). They have an extensive antimicrobial efficacy range according to the number, position and structure of substituent groups, the presence of glycosidic linkages alkylation of OH groups, and the geography and climate of the nation of origin. Generally, bioactive plant extracts contain complex mixtures of chemicals, and their synergistic action can give an improved effect (Wagner and Ulrich-Merzenich 2009). These compounds can damage microbial cells by affecting their cell membrane, DNA and protein synthesis, efflux pumps (EPs) in different ways (Khameneh 2019).

Even though there is a wide existence of these phytochemicals, based on their chemical composition, chemical structures, pathway of biosynthesis or solubility, they can be classified into some major groups that include alkaloids, sulfur-containing compounds, phenolic compounds, terpenes and coumarins (Kabera et al. 2014). The mechanisms of action and antimicrobial efficacy of the few vital compounds from those chemical groups are described below and given in Table 7.1. Different authors documented the chemical structure of many phytochemicals; some of those important chemical structures are given in Figure 7.1 (Cowan 1999; Antika, Triana and Ernawati 2020; Nakamoto et al. 2020).

7.5.1 Alkaloids

Alkaloids are chemically one of the most diverse structures of heterocyclic nitrogen compounds that possess antispasmodic, analgesic and antimicrobial efficacy. These compounds are frequently observed to play an important role in the treatment of a lot of infections (Stary 1994).

Indoquinoline alkaloids showed significant effects against yeast and Gram-negative bacteria (Cushnie et al. 2014). Alkaloid quinine is very well-known for its antiprotozoal efficiency against the parasite of malaria (Iwu, Angela and Okunji 1999; Antika, Triana and Ernawati 2020). Mainly alkaloids function through EP inhibitory activity. Berberine (isoquinoline alkaloid) gathers in cells driven by the membrane potential and targets gyrase, topoisomerase IV, RNA polymerase and inhibits nucleic acid synthesis (Yi et al. 2007).

7.5.2 Sulfur-containing Compounds

Sulfur-containing compounds exhibited antiviral, antibacterial, antifungal, and antiprotozoal activities that are obtained from medicinal plants (Kim, Kubec and Musah 2006; Rehman and Mairaj 2013). The crucial compounds are ajoene, allicin, isothiocyanates and sulforaphane. These compounds showed effective results against both Gram-positive and Gram-negative bacteria (Nakamoto et al. 2020). Antimicrobial mechanisms of these compounds may include inhibition of the DNA and protein synthesis, and inhibition of sulfhydryl-dependent enzymes (Wu et al. 2012; Reiter et al. 2017). A few compounds can also rupture the cell wall structure and result in the outflow of cellular metabolites (Lin et al. 2000).

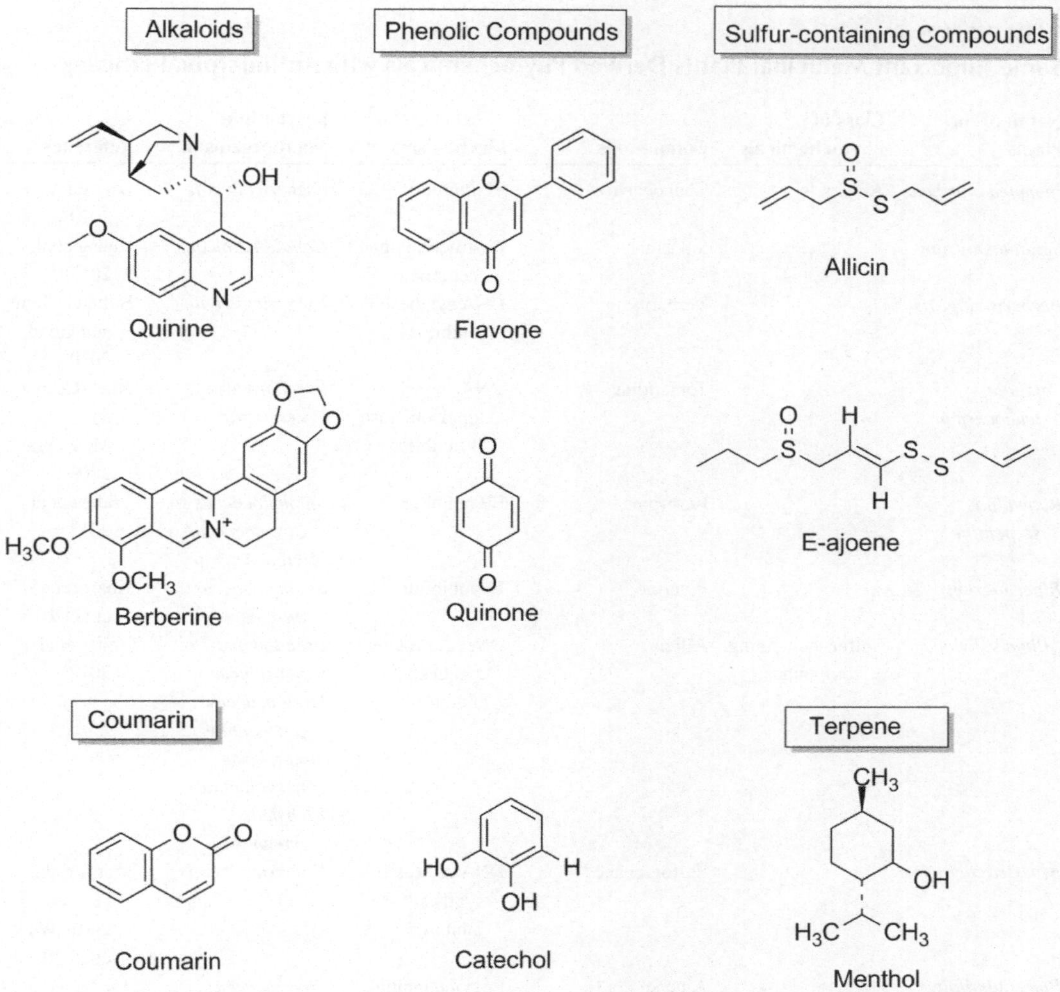

FIGURE 7.1 Chemical structure of some important antimicrobial compounds.

7.5.3 Phenolic Compounds

Phenolic compounds are a large diverse group of secondary metabolites obtained from medicinal plants. These compounds are broadly used against different pathogenic bacteria. These compounds kill various microorganisms by following mechanisms such as EP inhibition, modifying cell membrane permeability, rupturing cell membrane, DNA gyrase and different important enzyme inhibition (Li et al. 2006; Wu et al. 2008; Patel et al. 2015; Brown et al. 2015). These compounds contain flavonoids, flavones, flavanols, tannins and quinones. Flavonoids are famous for their antiviral, antimicrobial and anti-inflammatory activity (Cushnie and Lamb 2005; Gorniak, Bartoszewski and Kroliczewski 2019).

7.5.4 Terpenes

Terpenes are considered one of the most diverse groups of phytochemicals that perform various cellular functions (Oldfield and Lin 2012). Terpenes are mainly responsible for the peculiar odour of

TABLE 7.1
Some Important Medicinal Plants Derived Phytochemicals with Antimicrobial Efficacy

Name of the Plants	Class of Phytochemicals	Compounds	Mechanisms	Susceptible Microorganisms	Reference
Ipomoea muricata	Alkaloids	Chanoclavine	EP inhibitor	*Escherichia coli*	Dwivedi et al. 2019
Solanum nigrum		Solasidine	Destruction of cell membrane	*Candida albicans*	Chang et al. 2017
Berberis vulgaris		Berberine	DNA synthesis inhibitor	*Escherichia coli*	Boberek, Jem and Good 2010
Solanum leucocarpum		Tomatidine	DNA topoisomerase II inhibitor	*Saccharomyces cerevisiae*	Nino, Correa and Mosquera 2009
Rauwolfia serpentine		Reserpine	EP inhibitor	*Staphylococcus sp.* *Streptococcus sp.* *Micrococcus sp.*	Sridevi et al. 2017
Piper nigrum		Piperine	EP inhibitor	*Staphylococcus aureus*	Khameneh et al. 2015
Allium sativum	Sulfur- containing compounds	Allicin	DNA & protein synthesis inhibitor	*Pseudomonas aeruginosa* *Streptococcus pneumonia* *Streptococcus dysgalactiae* *Klebsiella pneumonia*	Reiter et al. 2017
Brassica oleracea		Sulforaphane	DNA & protein synthesis inhibitor	*Escherichia coli*	Matusheski and Jeffery 2001; Wu et al. 2012
Rubus ulmifolius		Ajoene	Enzyme inhibitor	*Streptococcus pneumoniae*	Talekar et al. 2014
Alpinia calcarata	Phenolic compounds	Kaempferol Galangin Pinocembrin	EP inhibitor	*Staphylococcus aureus*	Randhawa et al. 2016
Terminalia chebula		Chebulinic acid	DNA gyrase inhibitor	*Mycobactrium tuberculosis*	Patel et al. 2015
Hydrastis canadensis		Luteolin	EP inhibitor	*Staphylococcus aureus*	Brown et al. 2015
Camellia sinensis		Epigallocatechin gallate	β-ketoacyl-reductase inhibitor	*Escherichia coli*	Li et al. 2006
Cedrus deodara		Apigenin	d-alanine:d-alanine ligase inhibitor	*Escherichia coli* *Helicobacter pylori*	Wu et al. 2008
Origanum vulgare	Terpenes	Thymol	Cell membrane ATPase inhibitor	*Candida glabrata* *Candida albicans* *Candida krusei*	Sharifzadeh et al. 2018
Melaleuca alternifolia		Dehydroabietic acid	Destruction of cell membrane	*Staphylococcus aureus*	Soderberg et al. 1996

TABLE 7.1 (Continued)
Some Important Medicinal Plants Derived Phytochemicals with Antimicrobial Efficacy

Name of the Plants	Class of Phytochemicals	Compounds	Mechanisms	Susceptible Microorganisms	Reference
Podocarpus nagi		Totarol	respiratory inhibition	*Pseudomonas aeruginosa*	Haraguchi et al. 1996
Ferulago campestris	Coumarins	Aegelinol	DNA gyrase inhibitor	*Staphylococcus aureus* *Salmonella thypii*, *Enterobacter cloacae* *Enterobacter earogenes*	Basile et al. 2009
Mesua ferrea		Bergamottin epoxide	EP inhibitor	*Staphylococcus aureus*	Roy et al. 2013
Prangos hulusii		4′-Senecioiloxyosthol	DNA gyrase inhibitor	*Staphylococcus aureus Bacillus subtilis* *Klebsiella pneumoniae*	Tan et al. 2017

plants (Paduch et al. 2007). Terpenes derivatives are monoterpens, diterpenoids, etc. They are lipophilic in nature and assist in penetration into the cell walls of microbes (Copp 2003). Monoterpenes usually function by modifying membrane permeability and fluidity, changing the topology of its protein structure and creating disruption in the respiration chain (Oldfield and Lin 2012). These compounds exhibited antimicrobial activity against protozoa, viruses, bacteria and fungi (Haraguchi et al. 1996; Basile et al. 2009; Zuzarte et al. 2011; Chouhan, Sharma and Guleria 2017).

7.5.5 Coumarins

Coumarins possess significant antimicrobial efficacy obtained from medicinal plants (Smyth, Ramachandran and Smyth 2009; Roy et al. 2013; Tan et al. 2017). These compounds can repress the quorum sensing network of bacterial pathogens, i.e., the formation of little signal molecules by bacterial cells that disturb the development of biofilm formation and virulence factor formation (Reen, Gutierrez-Barranquero and Parages 2018; Zhang et al. 2018). Furthermore, some coumarins showed inhibitory action on the EP system and DNA gyrase (Roy et al. 2013; Tan et al. 2017).

7.6 INTERACTION BETWEEN MEDICINAL PLANT-DERIVED PHYTOCHEMICALS AND CONVENTIONAL ANTIBIOTICS

Many researchers documented the synergistic effect of phytochemicals and conventional antibiotics that enhance the activity of conventional antibiotics (Vaou et al. 2021). Antibiotic β-lactams together with α-mangostin, which is isolated from mangosteen (*Garcinia mangostana*) fruit, significantly enhance the activity of the treatment of β-lactam resistant bacterial strains. It is probable that together these components may inhibit the bacterial β-lactamase enzyme, therefore reactivating the antibiotic (Phitaktim et al. 2016). Numerous in vitro studies have documented the combination of plant extracts with antibiotics significantly reducing the minimum inhibitory concentration (MIC) of the antibiotics against a number of resistant strains and also resistance modifying activity (RMA) (Gibbons et al. 2003; Al-hebshi, Mohammed and Skaug 2006). Antibiotic-plant extract interaction depends on numerous factors, including pharmacodynamics and pharmacokinetics (Szalek, Grzeskowiak and Kozielczyk 2006). Pharmacodynamic plant extract-antibiotic interactions occur generally as synergism, antagonist and additive effects. Pharmacokinetic interactions happen mostly by enhancing the permeability of antibiotics to the cell membrane of bacteria or by suppressing or inducing antibiotic-metabolizing enzymes and transporters, that harmfully affect the excretion, distribution, absorption and metabolism of parallel administered antibiotics (Aiyegoro, Afolayan and Okoh 2009).

Pyridine isolated from *Jatropha elliptica*, at a concentration of 75 μg/ml, was exhibited to enhance by 4-times the efficacy of norfloxacin and ciprofloxacin against Nor A expressing *Staphylococcus aureus* when experienced at sub-inhibitory concentrations (Marquez et al. 2005). Dried leaf extract of *Camellia sinensis* and nalidixic acid together enhance the inhibition effect against *Salmonella typhi*. Together (Cextract = 0.62 mg/ml) and nalidixic acid exhibited a MIC value that was 8 times lower (32 μg/ml) than when used singly (256 μg/ml) (Farooqui et al. 2015). Also, Isoflavones derived from the plant *Lupinus argentens* enriched the activity of the antibiotics berberine and norfloxacin (Morel et al. 2003). Different researchers also tested fungi for synergism between plant extract and antifungal compounds. Together, *Pelargonium graveolens* essential oil and amphotericin B plus ketoconazole exhibited synergism in strains of *Aspergillus sp.* strains and *Agastache rugosa* essential oil and ketoconazole against *Blastischizomyces capitatus* (Shin and Kang 2003; Seungwon 2003). Plant extracts and antibiotics together also can perform as resistance-modifying or modulating agents (Chusri et al. 2009).

7.7 FUTURE PROSPECTS

There is plenty of evidence suggesting that medicinal plant-derived phytochemicals are very efficient as antimicrobial agents. Medicinal plants are the source of countless novel antimicrobial agents and only a small percentage of them has been discovered. They are cheap, easily available and have few or no side effects. A few phytochemicals exhibited both antibacterial efficacy and antibiotic resistance modifying efficacy. A number of them, when together with antibiotics, can defeat antibiotic resistance in bacteria. Green synthesis nanoparticles obtained from different medicinal plant extracts along with different metal ions (copper, gold, silver, platinum, etc.) showed effective antimicrobial activity. The translation of in vitro studies to in vivo tests and ultimately into human clinical trials may be an important step in the development of new antimicrobial drugs. Detailed study of toxicity tests, interplay with other substances and mode of action of the isolated phytochemicals are also needed. Nevertheless, continued research should be necessary for a better understanding of the accurate mechanisms and also the pharmacokinetic and pharmacodynamic properties of the phytochemicals.

REFERENCES

Aiyegoro, O. A., Afolayan, A. J., and A. I. Okoh. "In vitro antibacterial activities of crude extracts of the leaves of *Helichrysum longifolium* in combination with selected antibiotics." *Afr. J. Pharm. Pharmacol.*3, no. 6 (2009): 293–300.

Al-hebshi, N., Mohammed, A. H., and N. Skaug. "In vitro antimicrobial and resistance-modifying activities of aqueous crude khat extracts against oral microorganisms." *Arch. Oral Biol.* 51, no. 3 (2006): 183–188.

Ali, M. A., Temoor, A., and W. Wu et al. "Advancements in plant and microbe-based synthesis of metallic nanoparticles and their antimicrobial activity against plant pathogens." *Nanomaterials.*10, no. 6 (2020): 1146.

Anthony, M. "DNA gyrase as a drug target." *Trends. Microbiol.*5, no. 3 (1997): 102–109.

Antika, L. D., Triana, D., and T. Ernawati. "Antimicrobial activity of quinine derivatives against human pathogenic bacteria." In *IOP Conference Series: Earth and Environmental Science*, vol. 462, no. 1, p. 012006. IOP Publishing, 2020.

Atri, A., Mosaab, E., and A. Bouzidi et al. "Green synthesis of copper oxide nanoparticles using *Ephedra Alata* plant extract and a study of their antifungal, antibacterial activity and photocatalytic performance under sunlight." *Heliyon.*9, no. 2 (2023)L e13484.

Basile, A., Sergio, S., and V. Spadaro et al. "Antimicrobial and antioxidant activities of coumarins from the roots of *Ferulago campestris* (Apiaceae)." *Molecules.*14, no. 3 (2009): 939–952.

Baym, M, Laura, K. S., and R. Kishony. "Multidrug evolutionary strategies to reverse antibiotic resistance." *Science.*351, no. 6268 (2016): aad3292.

Bazzaz, B. S. F., Sahar, S., Bahman, K., and H. Hosseinzadeh. "Effect of catechins, green tea extract and methylxanthines in combination with gentamicin against *Staphylococcus aureus* and *Pseudomonas aeruginosa*:-combination therapy against resistant bacteria." *J Pharmacopuncture.*19, no. 4 (2016): 312.

Bazzaz, B. S. F., Bahman, K., Mohammad, R. Z. O., and H. Hosseinzadeh. "In vitro evaluation of antibacterial activity of verbascoside, lemon verbena extract and caffeine in combination with gentamicin against drug-resistant *Staphylococcus aureus* and *Escherichia coli* clinical isolates." *Avicenna J. Phytomed.* 8, no. 3 (2018): 246.

Bereksi, M. S., Hafida, H., Chahrazed, B., and D. E. Abdelouahid. "Evaluation of antibacterial activity of some medicinal plants extracts commonly used in Algerian traditional medicine against some pathogenic bacteria." *Pharmacogn. J.* 10, no. 3 (2018): 507–512.

Boberek, J. M., Jem, S., and L. Good. "Genetic evidence for inhibition of bacterial division protein FtsZ by berberine." *PloS One.*5, no. 10 (2010): e13745.

Brown, A. R., Keivan, A. E., and D. Todd et al. "A mass spectrometry-based assay for improved quantitative measurements of efflux pump inhibition." *PLoS One.*10, no. 5 (2015): e0124814.

Bussiere, D. E., Muchmore, S. W., and G Christopher et al. "Crystal structure of ErmC', an rRNA methyltransferase which mediates antibiotic resistance in bacteria." *Biochem.*37, no. 20 (1998): 7103–7112.

Chang, W., Ying, L., Ming, Z., Sha, Z., Yan, L., and H. Lou. "Solasodine-3-O-β-d-glucopyranoside kills *Candida albicans* by disrupting the intracellular vacuole." *Food Chem. Toxicol.* 106 (2017): 139–146.

Chouhan, S., Sharma, K., and S. Guleria. "Antimicrobial activity of some essential oils-present status and future perspectives." *Medicines.*4, no. 3 (2017): 58.

Christenhusz, M. J., and J. W. Byng. "The number of known plants species in the world and its annual increase." *Phytotaxa.*261, no. 3 (2016): 201–217.

Christopher, W. "Molecular mechanisms that confer antibacterial drug resistance." *Nature.*406, no. 6797 (2000): 775–781.

Christopher, W. "Where will new antibiotics come from?." *Nat. Rev. Microbiol.* 1, no. 1 (2003): 65–70.

Chusri, S., Villanueva, I., Voravuthikunchai, S. P., and J. Davies. "Enhancing antibiotic activity: A strategy to control *Acinetobacter* infections." *J. Antimicrob. Chemother.* 64, no. 6 (2009): 1203–1211.

Copp, B. R. "Antimycobacterial natural products." *Nat. Prod. Rep.* 20, no. 6 (2003): 535–557.

Cowan, M. M. "Plant products as antimicrobial agents." *Clin. Microbial. Rev.*12, no. 4 (1999): 564–582.

Cushnie, T. P. T., and A. J. Lamb. "Antimicrobial activity of flavonoids." *Int. J. Antimicrob. Agents.* 26, no. 5 (2005): 343–356.

Cushnie, T. P. T., Benjamart, C., and A. J. Lamb. "Alkaloids: An overview of their antibacterial, antibiotic-enhancing and antivirulence activities." *Int. J. Antimicrob. Agents.* 44, no. 5 (2014): 377–386.

Davies, J., and D. Davies. "Origins and evolution of antibiotic resistance." *Microbiol. Mol. Biol.* 74, no. 3 (2010): 417–433.

Dib, K., OumKeltoum, E., Katim, A., Yahia, C., and A. Filali-Maltouf. "Antibacterial activity of plant extracts against periodontal pathogens: A systematic review." *J. Herb. Med.* 29 (2021): 100493.

Dwivedi, G. R., Maurya, A., and D. K. Yadav et al. "Synergy of clavine alkaloid 'chanoclavine' with tetracycline against multi-drug-resistant *E. coli*." *J. Biomol. Struct. Dyn.*37, no. 5 (2019): 1307–1325.

Dzotam, J. K., and V. Kuete. "Antibacterial and antibiotic-modifying activity of methanol extracts from six Cameroonian food plants against multidrug-resistant enteric bacteria." *Biomed Res. Int.* 2017 (2017): 1–19.

Farooqui, A., Khan, A., Borghetto, I., Kazmi, S. U., Rubino, S., and B. Paglietti. "Synergistic antimicrobial activity of *Camellia sinensis* and *Juglans regia* against multidrug-resistant bacteria." *PloS One.*10, no. 2 (2015): e0118431.

Ghosh, A., Bidus, K. D., Arup, R., Biplab, M., and G. Chandra. "Antibacterial activity of some medicinal plant extracts." *J. Nat. Med.*62, no. 2 (2008): 259–262.

Gibbons, S., Moyosoluwa, O., Nigel, C. V., and A. I. Gray. "Bacterial resistance modifying agents from *Lycopus europaeus*." *Phytochemistry*.62, no. 1 (2003): 83–87.

Górniak, I., Bartoszewski, R., and J. Króliczewski. "Comprehensive review of antimicrobial activities of plant flavonoids." *Phytochem. Rev.* 18, no. 1 (2019): 241–272.

Haraguchi, H., Shingo, O., Hisashi, M., and I. Kubo. "Mode of antibacterial action of totarol, a diterpene from *Podocarpus nagi*." *Planta Med.* 62, no. 2 (1996): 122–125.

Hemdan, B. A., Ahmed, M., and M. M.Elbatanony et al. "Bioactive *Azadirachta indica* and *Melia azedarach* leaves extracts with anti-SARS-CoV-2 and antibacterial activities." *Plos. One*.18, no. 3 (2023): e0282729.

Iwu, M. W., Angela, R. D., and C. O.Okunji. "New antimicrobials of plant origin." *Perspectives on new crops and new uses.* ASHS Press, Alexandria, VA 457 (1999): 462.

James, C. E., and M. Winterhalter. "The porin and the permeating antibiotic: a selective diffusion barrier in Gram-negative bacteria." *Nat. Rev. Microbiol.*6, no. 12 (2008): 893–903.

Jeevanantham, V., Tamilselvi, D., Bavaji, S. R., and S. Mohan. "Green formulation of gold nanoparticles and their antioxidative assays, antimicrobial activity and photocatalytic colour decay." *Bull. Mater. Sci.* 46, no. 1 (2023): 32.

Jin, L., Chunshan, Q., Xiyan, H., and S. Fan. "Potential pharmacological resources: Natural bioactive compounds from marine-derived fungi." *Mar. Drugs*.14, no. 4 (2016): 76.

Joshi, R. K. "Aperspective on the phytopharmaceuticals responsible for the therapeutic applications." In Recent Advances in Drug Delivery Technology, pp. 229–262. IGI Global, 2017.

Kabera, J. N., Edmond, S., Ally, R. M., and X. He. "Plant secondary metabolites: biosynthesis, classification, function and pharmacological properties." *J. Pharm. Pharmacol.*2, no. 7 (2014): 377–392.

Khameneh, B., Milad, I., Morteza, G., Davod, G. A., Bibi, S. F. B, and M. Iranshahi. "Investigation of the antibacterial activity and efflux pump inhibitory effect of co-loaded piperine and gentamicin nanoliposomes in methicillin-resistant *Staphylococcus aureus*." *Drug Dev. Ind. Pharm.*41, no. 6 (2015): 989–994.

Khameneh, B., Roudayna, D., Kiarash, G., and B. S. F. Bazzaz. "Breakthroughs in bacterial resistance mechanisms and the potential ways to combat them." *Microb. Pathogen.*95 (2016): 32–42.

Khameneh, B., Milad, I., Vahid, S., and B. S. F. Bazzaz. "Review on plant antimicrobials: a mechanistic viewpoint." *Antimicrob. Resist. Infect. Control.* 8, no. 1 (2019): 1–28.

Kim, S., Kubec, R., and R. A. Musah. "Antibacterial and antifungal activity of sulfur-containing compounds from *Petiveria alliacea* L." *J. Ethnopharmacol.*104, no. 1–2 (2006): 188–192.

Kursa, W., Agnieszka, J., Jakub, W, and R. Kowalski. "Antifungal effect of plant extracts on the growth of the cereal pathogen *fusarium spp.*- an in vitro study." *Agron. J.*12, no. 12 (2022): 3204.

Levy, S. B. "Active efflux mechanisms for antimicrobial resistance." *Antimicrob. Agents. Chemother.* 36, no. 4 (1992): 695–703.

Li, B. H.,Zhang, R., Du, Y. T, Sun, Y. H., and W. X.Tian. "Inactivation mechanism of the β-ketoacyl-[acyl carrier protein] reductase of bacterial type-II fatty acid synthase by epigallocatechin gallate." *Biochem. Cell Biol.* 84, no. 5 (2006): 755–762.

Lin, C. M., PrestonIII, J. F., and C. Wei. "Antibacterial mechanism of allyl isothiocyanate." *J. Food Prot.* 63, no. 6 (2000): 727–734.

Lomovskaya, O., Mark, S. W., and A. Lee et al. "Identification and characterization of inhibitors of multidrug resistance efflux pumps in *Pseudomonas aeruginosa*: novel agents for combination therapy." *Antimicrob. Agents. Chemother.*45, no. 1 (2001): 105–116.

Magiorakos, A. P., Srinivasan, A., and B Roberta et al. "Multidrug-resistant, extensively drug-resistant and pandrug-resistant bacteria: An international expert proposal for interim standard definitions for acquired resistance." *Clin. Microbiol. Infect.* 18, no. 3 (2012): 268–281.

Mahamoud, A., Jacqueline, C., Sandrine, A. F, Winfried, V. K., and J. M. Pages. "Antibiotic efflux pumps in Gram-negative bacteria: The inhibitor response strategy." *J. Antimicrob. Chemother.*59, no. 6 (2007): 1223–1229.

Manandhar, S., Shisir, L., and R. K. Dahal. "In vitro antimicrobial activity of some medicinal plants against human pathogenic bacteria."*J. Trop. Med.*2019 (2019): 1–5.

Marquez, B., Neuville, L., and N. J. Moreau. et al., "Multidrug resistance reversal agent from *Jatropha elliptica*." *Phytochem.*66, no. 15 (2005): 1804–1811.

Matusheski, N. V., and E. H. Jeffery. "Comparison of the bioactivity of two glucoraphanin hydrolysis products found in broccoli, sulforaphane and sulforaphane nitrile." *J. Agric. Food Chem.* 49, no. 12 (2001): 5743–5749.

MNPS. Medicinal Plant Names Services (MNPS). (2023). Available online: www.kew.org/science/our-science/science-servicesmedicinal-plant-names-services(accessed on 07 July 2023)

Mohammed, F. S.,Eylem, K., Mustafa, S., Muhittin, D., and M. Pehlivan. "Antioxidant and antimicrobial activities of ethanol extract of *Helianthemum salicifolium* (Cistaceae)." *Indian J. Nat. Prod. Resour.* 12, no. 3 (2021): 459–462.

Moloney, M. G. "Natural products as a source for novel antibiotics." *Trends. Pharmacol. Sci.*37, no. 8 (2016): 689–701.

Morel, C., Stermitz, F. R., Tegos, G., and K. Lewis. "Isoflavones as potentiators of antibacterial activity." *J. Agric. Food Chem.*51, no. 19 (2003): 5677–5679.

Motyka, S., Barbara, K., Halina, E., Izabela, K. G., and A. Szopa. "Comparative analysis of metabolic variations, antioxidant profiles and antimicrobial activity of *Salvia hispanica* (Chia) seed, sprout, leaf, flower, root and herb extracts." *Molecules.*28, no. 6 (2023): 2728.

Nakamoto, M., Kayo, K., Suzuki, J. I., and Y. Kodera. "Antimicrobial properties of hydrophobic compounds in garlic: Allicin, vinyldithiin, ajoene and diallyl polysulfides." *Exp. Ther. Med.* 19, no. 2 (2020): 1550–1553.

Niño, J., Correa, Y. M, and O. M.Mosquera. "Biological activities of steroidal alkaloids isolated from *Solanum leucocarpum*." *Pharm. Biol.* 47, no. 3 (2009): 255–259.

Obaid, R. F.,Nada, K. K. H., Samah, A. K.,Lubab, A. J. A., and A. T.Jalil. "Antibacterial activity, anti-adherence and anti-biofilm activities of plants extracts against *Aggregatibacter actinomycetemcomitans*: An in vitro study in Hilla City, Iraq." *Casp. J. Environ. Sci.* 20, no. 2 (2022): 367–372.

Oldfield, E., and F. Y. Lin. "Terpene biosynthesis: Modularity rules." *Angewandte Chemie International Edition* 51, no. 5 (2012): 1124–1137.

Olsen, I. "New promising β-lactamase inhibitors for clinical use." *Eur. J. Clin. Microbiol.*34, no. 7 (2015): 1303–1308.

Osman, K., Dimitrios, E., and C. Basavannacharya et al. "An antibacterial from *Hypericum acmosepalum* inhibits ATP-dependent MurE ligase from *Mycobacterium tuberculosis*." *Int. J. Antimicrob. Agents.* 39, no. 2 (2012): 124–129.

Paduch, R., Martyna, K. S., Mariusz, T., and J. Fiedurek. "Terpenes: substances useful in human healthcare." *Archivum Immunologiae et Therapiae Experimentalis* 55, no. 5 (2007): 315–327.

Palhares, R. M., Marcela, G. D., Bruno, S. A. F. B, Gustavo, P. C., Maria, G. L. B., and G. Oliveira. "Medicinal plants recommended by the world health organization: DNA barcode identification associated with chemical analyses guarantees their quality." *PLoS One.*10, no. 5 (2015): e0127866.

Patel, K., Chetna, T., and S. Goyal et al. "Identification of chebulinic acid as potent natural inhibitor of *M. tuberculosis* DNA gyrase and molecular insights into its binding mode of action." *Comput. Biol. Chem.* 59 (2015): 37–47.

Phitaktim, S., Chomnawang, M., Sirichaiwetchakoon, K., Dunkhunthod, B., Hobbs, G., and G. Eumkeb. "Synergism and the mechanism of action of the combination of α-mangostin isolated from *Garcinia mangostana* L. and oxacillin against an oxacillin-resistant *Staphylococcus saprophyticus*." *BMC Microbiol.* 16, no. 1 (2016): 1–14.

Prasad, R. "Synthesis of silver nanoparticles in photosynthetic plants." *Journal of Nanoparticles*. Article ID 963961, 2014 (2014). http://dx.doi.org/10.1155/2014/963961

Randhawa, H. K.,Hundal, K. K., Ahirrao, P. N., Jachak, S. M., and H. S.Nandanwar. "Efflux pump inhibitory activity of flavonoids isolated from *Alpinia calcarata* against methicillin-resistant *Staphylococcus aureus*." *Biologia.*71, no. 5 (2016): 484–493.

Reen, F. J., Gutiérrez-Barranquero, J. A., and M. L. Parages. "Coumarin: A novel player in microbial quorum sensing and biofilm formation inhibition." *Appl. Microbiol. Biotechnol.* 102, no. 5 (2018): 2063–2073.

Rehman, F., and S. A. M. Y. A. Mairaj. "Antimicrobial studies of allicin and ajoene." *Int. J. Pharm. Bio. Sci.*4, no. 2 (2013): 1095–1105.

Reichling, J., Paul, S., Ulrike, S., and R. Saller. "Essential oils of aromatic plants with antibacterial, antifungal, antiviral, and cytotoxic properties–an overview." *J. Complement. Med. Res.* 16, no. 2 (2009): 79–90.

Reiter, J., Natalja, L., Van der Linden, M., Martin, G., Christian, M., and A. J. Slusarenko. "Diallylthiosulfinate (Allicin), a volatile antimicrobial from garlic (*Allium sativum*), kills human lung pathogenic bacteria, including MDR strains, as a vapor." *Molecules* 22, no. 10 (2017): 1711.

Reker, D., Anna, M. P., and T. Rodrigues et al. "Revealing the macromolecular targets of complex natural products." *Nat. Chem.* 6, no. 12 (2014): 1072–1078.

Roy, S. K., Kumari, N., and S. Pahwa et al. "NorA efflux pump inhibitory activity of coumarins from *Mesua ferrea*." *Fitoterapia.*90 (2013): 140–150.

Ruddaraju, L. K., Pammi, V. K., GirijaS. G., Veerabhadra S. P., and V. R. M. Kolapalli. "A review on antibacterials to combat resistance: From ancient era of plants and metals to present and future perspectives of green nano technological combinations." *Asian J. Pharm. Sci.*15, no. 1 (2020): 42–59.

Saranraj, P., and S. Sivasakthi. "Medicinal plants and its antimicrobial properties: A review." *Glob. J. Pharmacol.*8, no. 3 (2014): 316–327.

Schneider, T., and H. G.Sahl. "An oldie but a goodie–cell wall biosynthesis as antibiotic target pathway." *Int. J. Med. Microbiol.* 300, no. 2–3 (2010): 161–169.

Seungwon, S. "Anti-*Aspergillus* activities of plant essential oils and their combination effects with ketoconazole or amphotericin B." *Arch. Pharm. Res.*26, no. 5 (2003): 389–393.

Sharifzadeh, A., Khosravi, A. R., Shokri, H., and H. Shirzadi. "Potential effect of 2-isopropyl-5-methylphenol (thymol) alone and in combination with fluconazole against clinical isolates of *Candida albicans*, *C. glabrata* and *C. krusei*." *J. Mycol. Med.* 28, no. 2 (2018): 294–299.

Shaw, K. J., Rather, P. N., Hare, R. S., and G. H. Miller. "Molecular genetics of aminoglycoside resistance genes and familial relationships of the aminoglycoside-modifying enzymes." *Microbiol. Rev.* 57, no. 1 (1993): 138–163.

Shin, S., and C. A. Kang. "Antifungal activity of the essential oil of *Agastache rugosa* Kuntze and its synergism with ketoconazole." *Lett. Appl. Microbiol.* 36, no. 2 (2003): 111–115.

Shiu, W. K. P., Rahman, M. M., and J. Curry et al. "Antibacterial acylphloroglucinols from *Hypericum olympicum*." *J. Nat. Prod.* 75, no. 3 (2012): 336–343.

Sienkiewicz, M., Monika, L., Julita, C., Pawel, D., and E. Kowalczyk. "Antibacterial activity of thyme and lavender essential oils." *Med. Chem.*7, no. 6 (2011): 674–689.

Singh, A., and G. Chandra. "Larvicidal, pupicidal, repellence and smoke toxic efficacies of *Nicotiana plumbaginifolia* leaves against *Culex vishnui* mosquito." *Int. J. Trop. Insect Sci.* 42, no. 2 (2022): 1187–1195.

Singh, A., Bhattacharya, K., and G. Chandra. "Larvicidal potentiality of the leaf extracts of *Glochidion lanceolarium* (Phyllanthaceae) against the Japanese encephalitis vector *Culex vishnui* (Culicidae)." *Orient. Insects.* 53, no. 2 (2019): 267–278.

Smyth, T., Ramachandran, V. N., and W. F. Smyth. "A study of the antimicrobial activity of selected naturally occurring and synthetic coumarins." *Int. J. Antimicrob. Agents.* 33, no. 5 (2009): 421–426.

Söderberg, T. A., Anders, J., and R. Gref. "Toxic effects of some conifer resin acids and tea tree oil on human epithelial and fibroblast cells." *Toxicology.*107, no. 2 (1996): 99–109.

Sridevi, D., Shankar, C., Prakash, P, Park, J. H., and K. Thamaraiselvi. "Inhibitory effects of reserpine against efflux pump activity of antibiotic resistance bacteria." *Chem. Biol. Lett.*4, no. 2 (2017): 69–72.

Starý, F. *The Natural Guide to Medicinal Herbs and Plants*. Barnes & Noble, 1994.

Steglińska, A., Anastasiia, B., and P. Wawrzyniak et al., "Antimicrobial activities of plant extracts against *Solanum tuberosum* L. Phytopathogens." *Molecules.*27, no. 5 (2022): 1579.

Szalek, E., Grzeskowiak, E., and J. Kozielczyk. "Activity of the scientific network" interactions between herbal and synthetic drugs. Advantages and risks" Pharmacokinetic implications of herb-drug interactions." *Herba Pol.* 52, no. 4 (2006).

Tahir, K., Sadia, N., and A. Ahmad et al. "Facile and green synthesis of phytochemicals capped platinum nanoparticles and in vitro their superior antibacterial activity." *J. Photochem. Photobiol. B: Biol.* 166 (2017): 246–251.

Tajkarimi, M. M., Salam, A. I., and D. O. Cliver. "Antimicrobial herb and spice compounds in food." *Food Control.*21, no. 9 (2010): 1199–1218.

Talekar, S. J.,Sopio, C.,Katie, N.,Keith, P. K.,Cassandra, L. Q., and J. E. Vidal. "220D-F2 from *Rubus ulmifolius* kills *Streptococcus pneumoniae* planktonic cells and *pneumococcal* biofilms." *PLoS One.*9, no. 5 (2014): e97314.

Tan, N., Seçil, Y. T., Merve, B., Emir, T., and M. Miski. "Antibacterial activities of pyrenylated coumarins from the roots of *Prangos hulusii*." *Molecules.*22, no. 7 (2017): 1098.

Tegos, G. P., Mark, H., and J. J. Strouse et al. "Microbial efflux pump inhibition: Tactics and strategies." *Curr. Pharm. Des.* 17, no. 13 (2011): 1291–1302.

Tenover, F. C. "Mechanisms of antimicrobial resistance in bacteria." *Am. J. Med.* 119, no. 6 (2006): S3–S10.

Termentzi, A., Nikolas, F., and A. L. Skaltsounis. "Natural resins and bioactive natural products thereof as potential anitimicrobial agents." *Curr. Pharm. Des.* 17, no. 13 (2011): 1267–1290.

Tlili, N., Walid, E., Ezzeddine, S., Abdelhamid, K., Saida, T., and N. Nasri. "The caper (*Capparis L.*): Ethnopharmacology, phytochemical and pharmacological properties." *Fitoterapia.*82, no. 2 (2011): 93–101.

Tortorella, E., Pietro, T., P. E. Fortunato et al. "Antibiotics from deep-sea microorganisms: Current discoveries and perspectives." *Mar. Drugs.*16, no. 10 (2018): 355.

Vaou, N., Elisavet, S., Chrysa, V., Christina, T., and E. Bezirtzoglou. "Towards advances in medicinal plant antimicrobial activity: A review study on challenges and future perspectives." *Microorganisms.*9, no. 10 (2021): 2041.

Verpoorte, R. "Exploration of nature's chemodiversity: The role of secondary metabolites as leads in drug development." *Drug Discov. Today.* 3, no. 5 (1998): 232–238.

Victor, K. "Potential of Cameroonian plants and derived products against microbial infections: A review." *Planta. Med.* 76, no. 14 (2010): 1479–1491.

Vijay, R., Drisya, V. M., and D. R. F. Selta et al. "Synthesis and characterization of silver nanomaterial from aqueous extract of *Commelina forskaolii* and its potential antimicrobial activity against Gram negative pathogens." *J. King Saud Univ. Sci.*35, no. 1 (2023): 102373.

Vila, J., Sara, M., and J. Sanchez-Céspedes. "Porins, efflux pumps and multidrug resistance in *Acinetobacter baumannii*." *J. Antimicrob. Chemother.* 59, no. 6 (2007): 1210–1215.

Wagner, H., and G. Ulrich-Merzenich. "Synergy research: approaching a new generation of phytopharmaceuticals." *Phytomedicine.*16, no. 2–3 (2009): 97–110.

Wangchuk, P., Paul, A. K., and S. G. Pyne et al. "Evaluation of an ethnopharmacologically selected Bhutanese medicinal plants for their major classes of phytochemicals and biological activities." *J. Ethnopharmacol.* 137, no. 1 (2011): 730–742.

Webber, M. A., and L. J. V. Piddock. "The importance of efflux pumps in bacterial antibiotic resistance." *J. Antimicrob. Chemother.* 51, no. 1 (2003): 9–11.

WHO. *Antimicrobial Resistance*,2015. Available online: https://apps.who.int/gb/ebwha/pdf_files/WHA68/A68_20-en.pdf(accessed on 08 August 2022)

Wu, D., Kong, Y., and C. Han et al. "D-Alanine: D-alanine ligase as a new target for the flavonoids quercetin and apigenin." *Int. J. Antimicrob. Agents.* 32, no. 5 (2008): 421–426.

Wu, H. Z.,Fei, H. J., Zhao, Y. L.,Liu, X. J., Huang, Y. J., and S. W. Wu. "Antibacterial mechanism of sulforaphane on *Escherichia coli*." *Sichuan da xue xue bao. Yi xue ban= Journal of Sichuan University. Medical Science Edition* 43, no. 3 (2012): 386–390.

Yadav, R. N. S., and M. Agarwala. "Phytochemical analysis of some medicinal plants." *J. Phytol.*3, no. 12 (2011).

Yi, Z. B., Yu, Y., Liang, Y. Z., and B. Zeng. "Evaluation of the antimicrobial mode of berberine by LC/ESI-MS combined with principal component analysis." *J. Pharm. Biomed. Anal.* 44, no. 1 (2007): 301–304.

Zhang, Y., Andrea, S., and H. V. Acker et al. "Coumarin reduces virulence and biofilm formation in *Pseudomonas aeruginosa* by affecting quorum sensing, type III secretion and C-di-GMP levels." *Front. Microbiol.* 9 (2018): 1952.

Zuzarte, M., Gonçalves, M. J., andC. Cavaleiro. et al. "Chemical composition and antifungal activity of the essential oils of *Lavandula viridis* L'Her." *J. Med. Microbiol.* 60, no. 5 (2011): 612–618.

8 Nanomaterials as Alternative Antimicrobial Agents in Management of Phytopathogens in Agriculture

Diksha Tajane, Hariprasad Paikrao, and Anita Patil

8.1 INTRODUCTION

Nanotechnology is a promising new field of science with numerous potential applications in biology, agriculture, and medicine, among others. The application of nanotechnology in agriculture science is advantageous, as it aids production and potential in agriculture and livestock. Nanoparticle synthesis and application in various fields are two of the most important aspects of nanotechnology. From a result-oriented standpoint, the synthesis of nanomaterials is a complex process requiring specialized skills and technologies. Characterization is an essential step in determining the best particle size for a specific size range (Khan and Rizvi, 2014). In 1974, Norio Taniguchi (Tokyo Science University) proposed the term "nanotechnology" for materials precision scale for manufacturing. The term "nano" is commonly used to refer to one billionth of a metre or 10^{-9} in length (Prasad and Shrivastav, 2014). Various plants and microorganisms are used in agricultural nanotechnology to form nanoparticles, aggregated in inorganic material either outside or inside the cell to form nanomaterials.

Metal nanomaterial biosynthesis necessitates the use of many microorganisms for the formation of metal nanoparticles. Bacteria, fungi, yeast, viruses, and actinomycetes are capable of biosynthesizing silver, gold, platinum, tellurium, selenium, palladium, platinum, zirconium, and silica nanoparticles, among other metals (Narayanan and Sakthivel, 2010). Silver nanoparticles (AgNPs) are the sole nanoparticles that have been used in agriculture among many others (Vance et al., 2015). In the manufacture of these nanoparticles, two approaches are used: the "bottom-up" approach and the "top-down" approach. The detection of manufacturing defects in nanoparticles is one application of the bottom-up approach (Husen and Jawaid, 2020).

Bacteria are used to sustain life during the process of metabolization of oxygen. In contrast, silver ions prevent the metabolization of oxygen and destroy the enzyme to encourage the continuation of life. Fungi and bacteria die due to the suffocation they experience (Alvarez-Puebla et al., 2004). The use of a "green synthesis" approach can help to reduce costs while also limiting the use of hazardous chemicals. This approach can also help create safe and environmentally friendly nanoparticles (Chaudhry et al., 2018).

The nanoparticles can be used to distribute agrochemicals for a variety of purposes. Carbon nanoparticles in carriers, the magnetic property of iron nanomaterials, and the porous nature of silica

DOI: 10.1201/9781003268895-8

nanomaterials in the encapsulating of agrochemicals such as pesticides, insecticides, and antibiotics are all great candidates for delivery agents (Anandhi et al., 2020; Selyutina et al., 2020, 2017).

8.2 PHYTOPATHOGENS RESISTANT TO CONVENTIONAL ANTIMICROBIALS

Antibiotic resistance against human diseases and farm animals is extensively studied. The resistance may occur due to mutations at the molecular level in pathogens (alteration of target site in protein) causing inactivation of the antibiotic mechanism (Raman et al., 2020). Drug resistance occurs through horizontal gene transfer, study of such antibiotic resistance genes in pathogen genomes is termed as antibiotic resistome (Wright, 2007). Several antibiotic resistance mechanisms were proposed in the earlier research, such as inhibition of antibiotic cellular entry into the target site and alteration of antimicrobial substances (Sundin and Wang, 2018).

Conventionally, various antibiotics were tested against plant diseases such as chloramphenicol, penicillin, aureomycin, streptomycin and oxytetracycline (Anderson and Gottlieb, 1952; Leben and Keitt, 1954). However, only oxytetracycline and streptomycin were more effectively used in plant disease control. The variety of plant pathogens such as *Pseudomonas syringae* (apple tree infection of flower and fruit), *Xanthomonas campestris* (tomato bacterial spot) *Erwinia amylovora* (apple tree fire blight) are managed by streptomycin (McManus et al., 2002), while bacterial spot disease of nectarine and peach caused due to *Xanthomonas arboricola* pv. pruni was primarily treated with oxytetracycline, moreover streptomycin-resistant fire blight disease of apple orchards in Michigan (United States) was managed using oxytetracycline as a secondary antibiotic (McManus and Jones, 1994; Moller et al., 1981). Due to the high cost of antibiotics, they are primarily used for disease management of ornamental, vegetable and fruit crops (McManus et al., 2002). In the present section, some of the conventional antimicrobials and their resistance in different plant pathogens is discussed,

8.2.1 STREPTOMYCIN

Streptomycin ($C_{21}H_{39}N_7O_{12}$) is a primary antibiotic treatment used in veterinary and human medicine; it was originally isolated from *streptomyces griseus*. In plant disease management, streptomycin alone or along with nitrate and sulphate conjugation is applied, e.g. streptomycin sulphate results due to the combination of sulphate anions of H_2SO_4 and streptomycin. It belongs to a class of aminoglycoside antibiotics, which specifically binds with ribosomes to stop protein synthesis (translation), resulting in the death of the bacteria (Carter et al., 2000). Due to the longest and most widespread application of streptomycin in agriculture, it became the likely candidate for the generation of resistance against several pathogens. *Erwinia amylovora* was the first bacteria observed as resistant against streptomycin in orchards the possible mechanism was proposed due to *str A* and *str B* two genes with transposon Tn5393 (McManus and Jones, 1994). Moreover, some other genes also exhibit streptomycin resistance such as *aadA* which encodes nucleotidyltransferase enzyme observed in *Xanthomonas*, *Campylobacter*, *Acinetobacter* and transposons (Tn7, Tn1401, Tn21, Tn2670), While *apb (6)-1a*, *1b* and *1c* encoding phosphotransferase enzyme in *Salmonella*, *Streptomyces*, *Citrobacter*, *Proteus* and transposon (Tn5), similarly *ant (3")* and (6") coding nucleotidyltransferase enzyme in *Lactococcus*, *Aeromonas*, *Providencia*, *Leclaria*, *Staphylococcus*, *Proteus* and transposon (Tn1826) (Hollingshead and Vapnek, 1985; Mazodier et al., 1985; Perreten et al., 1997; Projan et al., 1988; Srinivasan et al., 2008; Sundin et al., 2016; van Overbeek et al., 2002; Wiener et al., 1998). Although the increasing resistance of streptomycin is observed US environmental protection agency gave approval for the use of streptomycin for about 100,000 acres in Florida for the management of citrus greening, also known as Huanglongbing caused due to *Liberibacter* spp. The same approval is also challenged in court by farmer and environmentalist groups (Erickson, 2021).

8.2.2 Tetracycline

These are the broad-spectrum antibiotics employed against several Gram-negative and positive bacteria. Tetracyclines are derived from streptomyces species or produced semi-synthetically. The mechanism of action of tetracycline inhibits the attachment of aminoacyl tRNA to the ribosomal acceptor site, thus preventing the protein synthesis of the pathogen (Chopra and Roberts, 2001). Several derivatives of tetracyclines are produced from actinomycetes (demethylchlortetracycline) or synthetically (rolitetracycline, doxycycline, methacycline and lymecycline) (Clark et al., 2012; Cunha et al., 1982; Jarolmen et al., 1970; Nelson and Levy, 2011).

Tetracycline resistance in veterinary and human medicine is a much more alarming issue compared to resistance in plant disease management. Tetracycline resistance occurs in a variety of modes such as ribosomal protection of the target cell with the help of ribosome protection proteins encoded by *otrA* gene (Doyle et al., 1991), exiting the tetracycline molecule out of the cell to inhibit its activity with the help of efflux proteins encoded by *tetA* determinant (Guillaume et al., 2004), whereas in some cases interaction of tetracycline with rRNA binding sites are mutated thus inhibiting the attachment and resulting in tetracycline resistance (Nguyen et al., 2014), a unique mechanism where the tetracycline molecule is modified to an inactive form by target pathogens with the help of monooxygenase enzyme which hydroxylates tetracycline and results into the reduced affinity with ribosomes. The monooxygenase enzyme is encoded by *tetX* and *tet37* genes (Nguyen et al., 2014; Speer et al., 1991; Yang et al., 2004).

Tetracycline is not seen as very resistant to plant pathogens compared to streptomycin; about 138 *Erwinia amylovora* strains are susceptible to oxytetracycline, as reported in Pacific Northwest USA (Loper et al., 1991). *Agrobacterium tumefaciens* and *Pseudomonas syringae* were resistant to

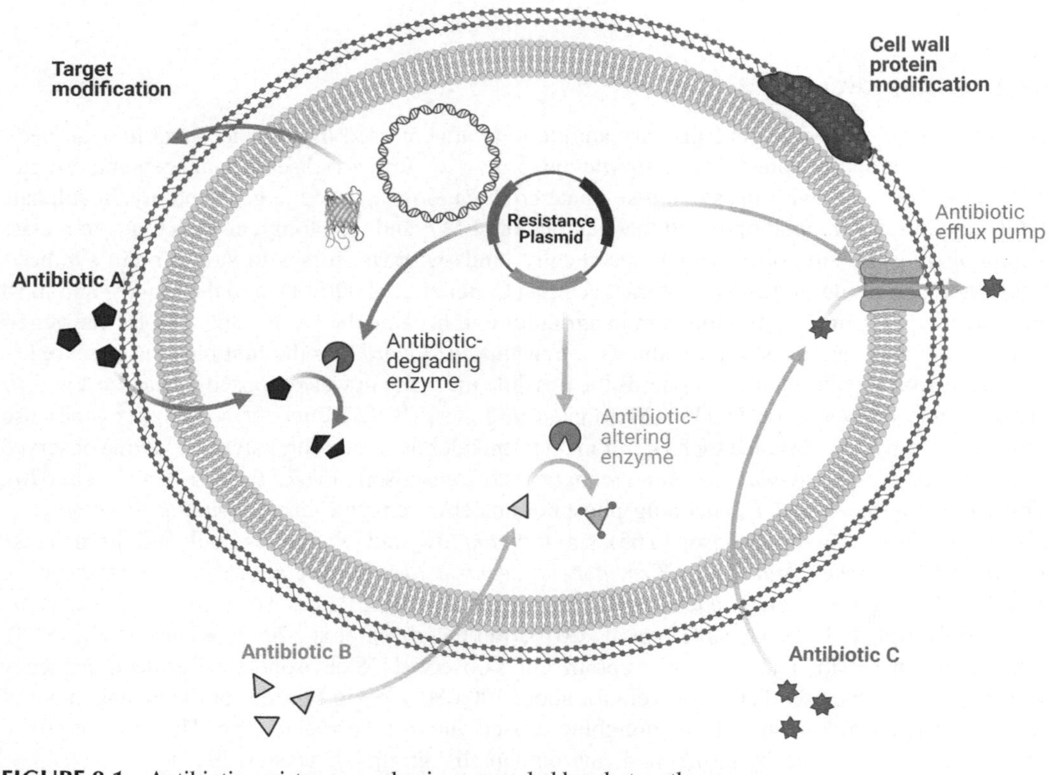

FIGURE 8.1 Antibiotic resistance mechanisms revealed by phytopathogens.

tetracycline in earlier reports (Hwang et al., 2005; Luo and Farrand, 1999; Spotts and Cervantes, 1995). Several plant pathogens possessed the efflux proteins of protein family TetA for tetracycline resistance such as *Xanthomonas perforans*, *Xanthomonas citri*, *Xanthomonas phaseoli*, *Pseudomonas aeruginosa*, *Pseudomonas syringae*, *P. fluorescence and P. putida*, *Erwinia piriflorinigrans and Ralstonia solancearum* (Sundin and Wang, 2018).

Apart from streptomycin and tetracycline being used as antibacteria, other antibiotics such as aureofungin, validamycin, kasugamycin, ningnanmycin and oxolinic acid (OA) are also used in plant disease management; they possess antiviral activity (Han et al., 2014), anti-water mould activity (Tso and Jeffrey, 1956) and antifungal potential (Lee et al., 2005; Vallad et al., 2010). Kasugamycin was primarily used in Japan against *Magnaporthe grisea*; also known as rice blast fungus it is also used for Japanese apricot, kiwi and *Beta vulgaris* in several countries (Spadafora et al., 2010). A pictorial illustration of the different resistance mechanisms of phytopathogens against antimicrobials is given in Figure 8.1. All the figures in the chapter are provided with permission of BioRender.com.

8.3 SYNTHESIS OF NANOPARTICLES

When it comes to the synthesis of nanomaterials, a number of physical, chemical, and biological techniques are applied.

8.3.1 Chemical Method

A chemical method is exercised commercially to make nanoparticles, and the method used varies depending on the materials used in the process. Some of these are briefly discussed below.

8.3.1.1 Solvothermal Deposition

A sufficient number of solvents at a specific temperature are used in a heterogeneous chemical reaction above the boiling point. The reaction takes place in a closed vessel (Byrappa and Yoshimura, 2012). Because of its extremely small grain size, water is used as a supercritical liquid, acting as both a liquid and a gas, lowering the surface tension and dissolving chemical compounds. These techniques are used to increase the solubility and reactivity of metals, among other things (Byrappa and Yoshimura, 2012). These hydrothermal conditions aid in forming Cu, Ge, Si, Pt, Au, and Ag nanoparticles. The effect of temperature variations and the amount of SDBS on the shape and size of copper nanoparticles is greater (Vincenzo Amendola et al., 2006). Recent studies confirm the one-step solvothermal synthesis of Ni nanoparticle catalysts embedded in ZrO_2 porous spheres (Meiliefiana et al., 2022).

8.3.1.2 Electrochemical Method

In this method, electricity is used for the production of nanoparticles from metals by passing the current between two electrodes. It is possible to avoid heat using a vacuum in this system (Raja et al., 2008).

This method is used because of its accessibility, as well as the fact that it has a very short reaction period, making it ideal for the fabrication of Cu nanoparticles. The copper nanoparticles produced by this method are around 10 nm (Khan and Rizvi, 2014). Microwave emissions are introduced into the reacting solution during the reaction (Komarneni, 2003). The frequency of 2.456 GHz is the most commonly used (Blosi et al., 2011). Because of the rapid and homogeneous nature of the reaction, this method is capable of producing extremely small nanoparticles. It was also possible to synthesize copper nanoparticles with the help of two different precursors and reducing agents, namely copper sulphate and sodium hypophosphite, respectively (Zhu et al., 2004). The silver nanoparticles were synthesized using an electrochemical method. The researchers used silver anode as a sacrificial source of Ag^+ while the rhamnolipid (surfactant from microbial source *Pseudomonas* sp.) (Hallmann and Macierzanka, 2017) is used for stabilization. Results indicated that increasing rhamnolipid concentration and temperature increases the formation of silver nanoparticles (AgNPs). The resultant AgNPs were tested against bacterial phytopathogens *Xanthomonas campestris* and *Agrobacterium tumefaciens*. Similarly, the phytotoxicity analysis showed that rhamnolipid reduces the phytotoxicity of AgNPs (Kuntyi et al., 2020).

8.3.1.3 Chemical Reduction Method

The synthesis of copper nanoparticles is executed under strict morphological control, and the control of particle size is possible through this procedure (Song et al., 2004). These methods for copper reduction use sodium borohydride as a reducing substance to reduce copper concentrations (Aslam et al., 2002). In a research study, silver nanoparticles were synthesized by the chemical reduction method (Fang et al., 2005). A very simple protocol was adopted for chemical synthesis where 50 ml solution $AgNO_3$ (10^{-3} M) was boiled and 1% tri-sodium citrate (5 ml) was added drop-wise till the colour change was observed. Later the solution was allowed to cool. Later, antibacterial applications of nanoparticles were evaluated against Gram-positive and Gram-negative bacteria. The results confirmed the antibacterial potency of chemically and biologically synthesized nanoparticles (Gudikandula and Charya Maringanti, 2016).

8.3.1.4 Microemulsion Technique

Microemulsions are produced by an amine-based co-surfactant solution that is homogeneous and clear. The formation of nanoparticles begins with thermodynamically stable dispersions from immiscible fluids such as oil in water or water in oil. In order to form a good emulsion, the ingredients of oil, water, and a surfactant must be combined (Chen et al., 2006). Sufficient surfactants are present to reduce the tension between water and oil, thereby facilitating the miscibility of oil and water (Kitchens et al., 2005).

Methods based on sonochemistry: The chemical reaction uses ultrasonic radiations with frequencies ranging from 10 to 20 kHz (Suslick et al., 1996). The ease with which this method operates, the controlled nanoparticles produced, and the simplicity of the procedure are the primary advantages of this method, which makes use of a precursor solution in various concentrations (Vijayakumar et al., 2000).

8.3.2 Biological Method

When key variables such as organism types, inheritability, and genetic qualities of organisms are taken into consideration, the biobased approach may be used to synthesize nanoparticles that are highly stable and well-characterized. Environmentally and economically friendly synthesis techniques that do not involve harmful compounds are becoming increasingly sought after in the industry. Scientists are now looking into the organisms as a result of this. Organisms that can synthesize nanoparticles range from fungi, and bacteria to plants (Mohanpuria et al., 2008; Prasad et al., 2016). Although physical and chemical methods are commonly used for nanoparticle synthesis, some drawbacks are observed, such as high energy requirements, heat generation, and expensive setup. Furthermore, toxic chemicals can pollute the environment (Soni et al., 2018). Thus, biological synthesis becomes a key method in nanoparticle generation. Several reports proposed different eco-friendly approaches to the green synthesis of nanoparticles using bacteria, fungi and plants, and later its applications in the agriculture and food sector (Bahrulolum et al., 2021; Sarlak and Abdi, 2022; Srivastava et al., 2021; Kisimba et al., 2023).

8.4 CHARACTERIZATION OF NANOPARTICLES CONJUGATED ANTIMICROBIALS

Synthesized nanoparticles need to be characterized by analytical techniques for their size, shape, charge, surface chemistry, morphology, area and dispersity. The current advancement in analytical techniques makes the characterization of antimicrobial nanoparticles easier and the methods can be divided into microscopic examination and molecular characterization.

8.4.1 Microscopic Examination of Nano-antimicrobials

Nanomaterials' functional information can be easily deduced from the microscopic characterization. The direct visualization of particle distribution and nature of nanoparticles decrypts the functional behaviour. The microscopy also reveals the antimicrobial activity of nanoparticles by providing particle-microbe interaction.

8.4.1.1 Electron Microscopic Analysis

Electron microscopy provides insights into nanoparticle-based antimicrobial materials due to its high-resolution capacity. It gives the shape, size and state of aggregation for nanoparticles (Sullivan et al., 2012). Although the destruction of samples during sample preparation and analysis occurs, these methods proved extensively useful in studying surface morphology. Some of the electron microscopy techniques used in nano-antimicrobial characterization are described below.

8.4.1.2 Scanning Electron Microscopy (SEM)

The principle of SEM lies in electron beam and specimen surface interaction. The surface molecules or atoms generate secondary electrons detected by photomultiplier through scintillation materials. However, the non-conducting materials cannot be analyzed easily, thus the nanomaterials are mostly coated with convicting elements such as carbon, chromium and gold (Nixon et al., 1971). SEM affords microorganisms and substrate interaction in detail. In a recent study bioaccumulation of zinc oxide nanoparticles in *Triticum aestivum* L. was assessed by SEM and ICP-MS method (Alsuwayyid et al., 2022).

8.4.1.3 Transmission Electron Microscopy (TEM)

TEM provides valuable information about the interactions between microbial cells and nanoparticles. TEM has been used to characterize the antimicrobial potential of silver nanoparticles in several studies. It has also been used to interpret the internalization of nanomaterials in microbial cell membranes (Choi et al., 2010). Researchers used carbon nanodots against drought stress in peanut plants and observed significant results, the microscopic characterization of the carbon nanodots was achieved by TEM, the size observed as 3.9 nm, while in another case salinity stress tolerance was observed in *Brassica napus* due to multiwalled carbon nanotubes (TEM size 6–12 nm) (Su et al., 2018; Zhao et al., 2019).

8.4.2 UV-visible Spectroscopic Analysis

UV-visible spectroscopy is a useful and quick tool in the characterization of nanoparticles because it provides a sharp absorption spectrum for a wide range of metal nanoparticles. Although it is not possible to gain precise relatable knowledge about the size distribution of nanomaterials using UV-visible spectroscopy, it can provide an indication of the shape of the nanoparticles (Liu et al., 2007; Sharma et al., 2009).

8.4.3 Infrared Spectroscopy

Infrared spectroscopy (IR) is one of the most effective non-destructive techniques for the characterization of nanoparticles. The infrared spectrum provides fingerprints of atomic vibrational frequencies for different molecules. In order to compare the selective metal nanoparticles used in antimicrobials or the conjugation of nanoparticles with biological materials such as enzymes, the

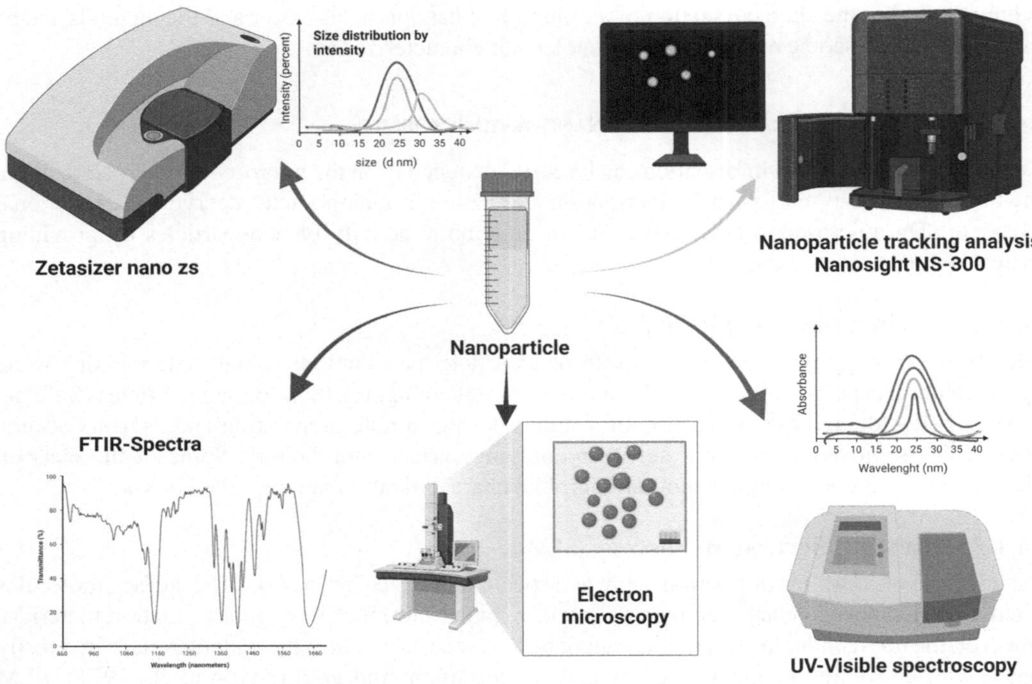

FIGURE 8.2 Nanoparticle's characterization methods.

internal standards can be used. Some advanced features that can be used in microbes-nanoparticle interactions include the Fourier transform, attenuated total reflectance, and infrared microscopy coupled to IR (Chongdar et al., 2005; Gunasekaran et al., 2004).

8.4.4 Dynamic Light Scattering Analysis

It is possible to determine the size distribution of nanoparticles in colloidal solutions by applying the principles of dynamic light scattering. Because the degree of repulsion between adjacent nanoparticles in a dispersion can be measured by using zeta potential, these techniques can be used to characterize the stability of solutions (Chongdar et al., 2005; Jaiswal et al., 2010).

Many other techniques, such as X-ray diffraction analysis, laser scanning confocal microscopy, scanning X-ray transmission microscopy, and energy-dispersive x-ray spectroscopy, have been mentioned to characterize antimicrobial nanoparticles by previous researchers (Sullivan et al., 2012) (Figure 8.2).

8.5 MECHANISM OF ACTION OF NANOPARTICLE ANTIMICROBIALS AGAINST PHYTOPATHOGENS

Plant pathogens include a variety of viruses, fungi, bacteria, and parasites that cause reduced plant productivity. To damage the plants and absorb their nutrients, pathogens release toxins, enzymes, and numerous secondary metabolites. Food poisoning, famines, massive financial losses, and the extinction of plant species are all caused by plant diseases (Agrios, 2009). Conventional agriculture employed a variety of troll methods, but as plant pathogen resistance has increased, new and more advanced methods must be implemented. Nanoscale particles and materials have demonstrated great promise in the fight against plant pathogens that are particularly resistant to antibiotics

(Khan et al., 2019). When it comes to plant protection, metal nanoparticles are commonly used. Silver nanoparticles are particularly popular because they are functional against a broad range of phytopathogens. The popularity of silver in the synthesis of nanoparticles can be attributed to the fact that it can be synthesized from a diversity of sources (biological, chemical and physical). The scientific community has long recognized the harmful effect of silver ions and silver-conjugated materials on microorganisms (Sondi and Salopek-Sondi, 2004).

Researchers proposed a variation of antimicrobial mechanisms of action for metal nanoparticles when they were applied to pathogens (Lemire et al., 2013). In phytopathogens, protein degradation has been observed differently, including the replacement of the active catalytic metal, the destruction of peptide bonds, and the modification of protein confirmations, among others. The nanoparticles generate free radicals in the form of reactive oxygen species, which can cause harm to bacterial and mitochondrial membranes, as well as impairment of membrane potential and, ultimately, death of the bacteria in some instances (Dhiman et al., 2022, 2023). Furthermore, the nanoparticles have the potential to cause double-stranded breaks in essential genes and inhibit the expression of nutrient transporter genes (Alghuthaymi et al., 2015; Lemire et al., 2013).

In the presence of opposite charges, the nanoparticles (positively charged) bind to the bacterial membrane (negatively charged), allowing them to move through and enter the cell. The pores are responsible for the expulsion of cell content (DNA, proteins, enzymes, and cytosol) as well as the cessation of respiratory activity. SEM and TEM analysis of bacteria confirmed the formation of pores and the entry of nanoparticles into the bacteria's cells (Gahlawat and Choudhury, 2019). Silver nanoparticles derived from biogenically synthesized silver exhibited significant antibacterial activity against the rice bacterial brown stripe pathogen *Acidovorax oryzae* strain RS-2. During infection, the pathogen forms a biofilm, which is disturbed by the Ag nanomaterials, and the cell membrane is damaged by the development of pits (Masum et al., 2019). In other studies, the antifungal activity of silver nanoparticles was observed against *Fusarium graminearum* in the form of hyphal distortion and cell wall damage, and similar results were observed against other fungal pathogens such as *Trichosporon ashii* and *Alternaria alternata*. The literature supported these results (Ibrahim et al., 2020; Xia et al., 2016).

When exposed to metal nanoparticles, they cause the manufacturing of reactive oxygen species, which deplete the activity of antioxidants and alter the permeability of cells. They also cause single and double-strand breaks in DNA, which ultimately cause genotoxicity in pathogenic bacteria (Aziz et al., 2014, 2015, 2016, 2019). Extensive research has been done to determine the effect of silver nanomaterials on *Azotobacter vinelandii* in various mechanisms such as cell injury, growth inhibition and oxidative stress. Results were obtained using electron spin resonance, transmission electron microscopy and flow cytometry (Zhang et al., 2018).

Nanoparticles can exert significant control over multidrug-resistant bacteria. Several nanoparticles carry nitic oxide, released when the particles come into contact with bacteria. Nitic oxide is an antimicrobial gaseous molecule that has found widespread use in biomedical applications against clinically significant multidrug isolates of bacteria such as *E.coli*, *Enterococcus faecalis*, and *Klebsiella pneumoniae*, among others (Hajipour et al., 2012). Agrochemicals are delivered to plants via nitric oxide nanoparticle carriers (NONc), which are being investigated for their anti-oxidative stress activity. Nitric oxide nanomaterials carriers (NONc) were investigated for their anti-oxidative stress activity. The anti-biofilm activity of NONc has been identified as an important mechanism of phytopathogen control (Seabra et al., 2014). A study investigates the antibacterial activity of magnesium oxide nanoparticles (MgONP) at odds with the phytopathogen *Ralstonia solanacearum*, which causes catastrophic tobacco wilt. The mechanism of action of MgONP involves cell membrane injury and the accumulation of reactive oxygen species, which ultimately ends in bacterial death (Cai et al., 2018). In a study of the silver nanoparticles-mediated response in pathogen-challenged (*Sclerotium rolfsii*) chickpea plants responsible for collar rot disease, it was discovered that the silver-nanoparticle-treated plants had altered lignification and H_2O_2 production, reduced sclerotia

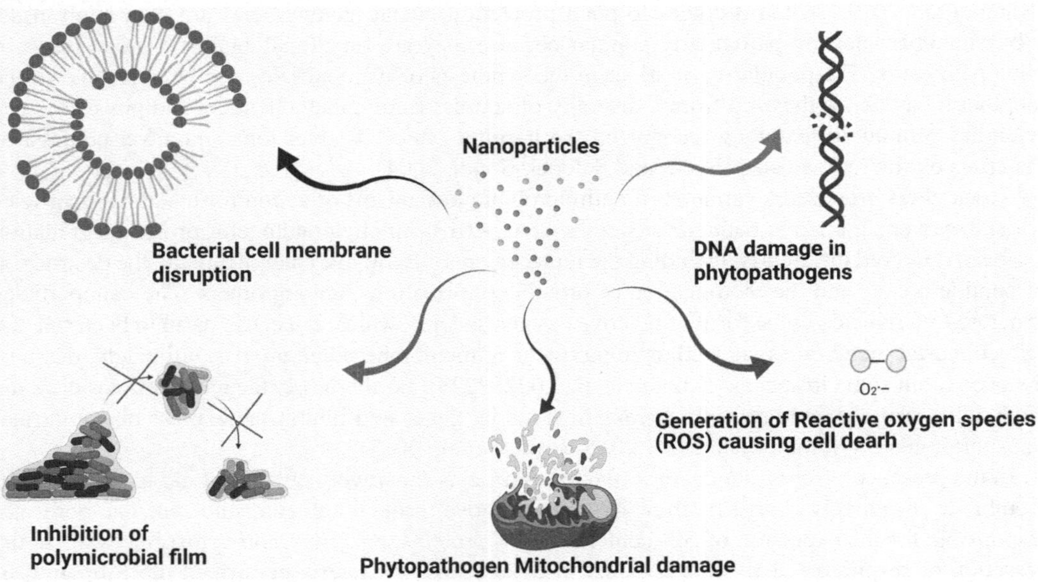

FIGURE 8.3 Mechanism of action employed by nanoparticles against plant pathogens.

germination, and increased the synthesis of phenolic acids when correlated with the pathogen challenged control plants (Mishra et al., 2017). The diagrammatic illustration of nanoparticles' antimicrobial mechanism is proposed in Figure 8.3.

8.6 TOXICITY OF METALLIC NANOPARTICLES ON PLANTS

Plant diseases can be managed through NPs, which can be applied to both the below-ground and above-ground parts of the plants. Pathogenic attacks and seedling diseases can be managed at a further stage by dipping the seeds into a solution of biosynthesized nanoparticles (NPs) (Tripathi et al., 2017). It is possible that the applied nanoparticles will stop the plant pathogens in a variety of plant parts. They will also transfer throughout the plant system, decreasing the occurrence of disease. On the other hand, plant toxicity was observed in diversified NPs as a result of their translocation, and accumulation within the plant. Metal nanoparticles (NPs) are rarely found in the environment, despite being known to be toxic, even at lower concentrations. Many factors influence the uptake and translocation of nanoparticles, including the characteristics of the NPs (composition, size, and surface area), the dose of application, the delivery technique, and the plant species. Plant physiology can be altered as a result of bioaccumulation, which can have an impact on the development and growth of plants. Most studies have found that excess metallic NPs in crop plants, such as *Solanum lycopersicum*, *Triticum aestivum*, *Allium cepa*, and *Cucurbita pepo*, can cause an oxidative burst by obstruction with the electron transport chain in these plants.

Additionally, it has the potential to impair the detoxification of reactive oxygen species, which could have genotoxic consequences. Even though research into the toxicity of nanoparticles in crops is still in its early stage, the application of new agro nanotechnology products and tools is crucial for the development of agriculture. As an outcome, the formation of phytohormones and secondary metabolites is inhibited, resulting in a reduction in plant growth rate. Some of the harmful effects and toxicity that have been observed in crops include restrictions in water transportation in plant systems, a decline in metabolic disorder, growth hormone production, chromosomal abnormalities, decreased growth, oxidative stress, deviation in the transcriptional profile of various genes, and

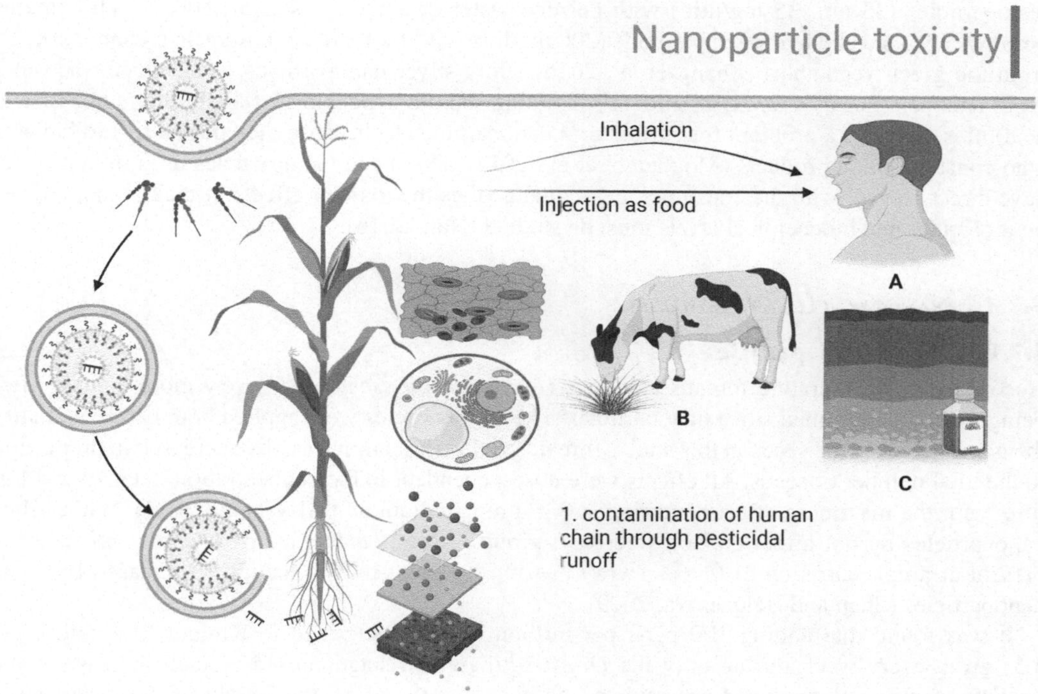

FIGURE 8.4 Toxicity of nanoparticles.

increased vulnerability to natural harmful materials, among other things (e.g., Arsenic) (Figure 8.4) (Jośko et al., 2017).

8.7 APPLICATIONS OF NANOPARTICLES IN AGRICULTURE

Consider a comparison between the delivery and absorption of nanoparticle-encapsulated and unbound medicines. In this situation, the delivery and uptake of medications contained in nanoparticles will be enhanced, as will their uptake by target cells. One will also notice that nanoparticle-encapsulated medications are less hazardous (De Jong and Borm, 2008). This approach can also be applied in the agricultural field. A further advantage of nanoparticles is that they have a highly reactive surface that makes them biologically active (Revell, 2006). This section will highlight attempts to use nanoparticles as fertilizer or crop plant stimulants, antibacterial agents and pesticide carriers, and antibacterial agents and pesticide carriers that have been reported. Because it is critical to contemplate the power of the negative consequences of using nanomaterials in agriculture, the phytotoxicity of nanoparticles will be briefly discussed in this section. Although biosensors and nanoparticles-mediated gene transfer in plants are also being investigated for agricultural applications, we will concentrate solely on nanomaterials in agricultural areas for nutrition and phytopathogen control in the most fundamental sense (Srilatha, 2011; Prasad 2014; Prasad et al., 2014; 2017).

It is important to remember that the application technique determines the path that nanoparticles take in their struggle to reach plants. By interacting with soil-associated nanoparticles, nanoparticles will be attracted to leaf and other atmospheric portions of plants. By interacting with airborne nanoparticles, nanoparticles will be attracted to roots. Following the application of nanoparticles to plants, there is an unexplored question of how to remove the nanoparticles that the plants have absorbed. For example, washing any green vegetables that had previously been treated with silver nanoparticles (35 nm, 35 mg/litre) with purified water or sanitizer tsunami 100, which contains peroxyacetic acid and hydrogen peroxide (200 mg/litre chlorine), did not eliminate all nanoparticles from the green vegetables (Zhang et al., 2016). Only silver nanoparticles were transformed into AgCl nanoparticles by Clorox, a sanitizer containing sodium hypochlorite (80 mg/litre peroxyacetic acid), this could be a problem for agricultural nanoparticle use because they pose a risk to humans who consume these products (Mogharabi et al., 2014). The metal nanoparticles used in plants can have direct contact with the consumer, e.g. humans; thus the toxicity effects of these nanoparticles on molecular and biochemical levels must be studied (Nair, 2016).

8.7.1 Nanoparticles as Fertilizers

8.7.1.1 Silica Nanoparticles

According to the literature tomato seedlings (*Lycopersicon esculentum*) grew more quickly after being treated with fumed silica nanoparticles. The nanoparticles were applied both before and after the germination of the seeds in this study. Tomato seed germination was also increased in proportion to the total number of seeds. All effects were dose-dependent in the concentration range of 2–14 g/litre, with the maximum effect occurring at 8 g/litre (Siddiqui and Al-Whaibi, 2014). While silica nanoparticles do not affect the soil microbial group or seed emergence of *Lactuca sativa*, when present at a concentration of 0.06% (w/w) in soil, the shoot-to-root length is increased by silica nanoparticles (Shah and Belozerova, 2009).

It was found that adding 100 parts per million (ppm) silicic acid to Kimura B solution (pH 5.5) given every week during only the *Oryza sativa* green stage caused a small increase in the number of spikelets produced per panicle, the height of the plant, the weight of the straw, and a 1.5-fold increase in the production of spikelet filled. If the solution had been applied continuously throughout the reproductive stage, the effect would have been more than 2.5 times as significant. It was discovered that the application of silicic acid during the ripening stage slightly accelerated the ripening process (Ma et al., 1989). Silica nanoparticles, due to their mesoporous nature, acquire most of the chlorophyll-protein complexes and thus increase the photosynthetic activity of the plant (Rajput et al., 2021; Verma et al., 2022). In one study silica nanoparticles were observed to reduce the plant's stress, thus considered as safe in delivery (Bhat et al., 2021). Similarly, deficiency of silica can result in poor growth and stress to the plants, thus silica nanoparticle-based fertilizers can prove a comprehensive booster for agriculture production (Adisa et al., 2019).

8.7.1.2 Selenium Nanoparticles

When applied to *Zea mays* seeds at a rate of 0.1 g/ha, selenium nanoparticles significantly increased crop production and the concentration of starch and protein in the kernels (Gudkov et al., 2020). A selenium fertilizer (10 g Se/kg, combined with calcium ammonium nitrate, muriate of potash, and triple superphosphate) applied at rates of 5–25 g Se/ha has no beneficial effect. It is unclear whether the difference in the second instance is due to the amount of selenium mixed with other components or whether the amount of selenium mixed with other components caused the difference in the first instance (Tremblay et al., 2015). Recent studies have proposed the role of selenium nanoparticles in agriculture and their behaviours in different environments such as air, plant, water and soil (Bano et al., 2021; Garza-García et al., 2022).

8.7.1.3 Gold Nanoparticles

When applied to soil at 0.013% (w/w), gold nanoparticles did not affect the microbial community or the seed germination of *Lactuca sativa*. However, they did increase the shoot-to-root length ration of the plant (Shah and Belozerova, 2009). The treatment of *Arabidopsis thaliana* seeds and seedlings with 80 g/ml gold nanoparticles led to a notable increase in plant growth. The treated plant bloomed earlier and produced longer pods than those produced by control plants. Several microRNAs (miRNAs) had their expression patterns altered as well. Despite an increase in antioxidant potential and antioxidant enzyme levels as a result of the implementation of nanoparticles, no evidence of oxidative stress was observed (Kumar et al., 2013). After being treated with gold nanoparticles, *Brassica juncea* seeds sprouted larger and grew more quickly. The amount of chlorophyll and sugar in the leaves, the number of seeds produced, the amount of seed oil produced, and the pods number produced per plant all increased. The number of secondary branches in the mustard crop treated with nanoparticles was higher than the number of primary branches.

8.7.1.4 Silver Nanoparticles

Brassica juncea seeds were treated with nanoparticles at concentrations of 25 and 50 parts per million (ppm). The treatment resulted in longer shoots and roots (Sharma et al., 2012). Silver nanoparticles at a 13.5 mg/litre concentration were found to increase the root length of *Raphanus sativus* seeds (Iram et al., 2014). The addition of 100 M silver nanoparticles to the growing medium of Arabidopsis plants in another study resulted in a stimulation of root development. Cell division, chloroplast growth, and glucose metabolism are all examples of processes in the body (Gallo et al., 2021). It was discovered that spraying *Borago officinalis* L. with silver nitrate and silver nanoparticles reduced seed abscission, increasing the seed numbers produced in the crop. Additionally, a wide range of other metrics, such as number of leaves, height, length, width, and dry weight of the plant, were increased. The amount of silver nanoparticles required to meet the requirement was less than the amount of silver nitrate required to achieve the desired result. The total silver absorption by *Borago officinalis* was discovered to be concentration-dependent in the range of 20–60 ppm of nanoparticles and 100–300 ppm of $AgNO_3$ in the presence of nanoparticles (Sah et al., 2011).

8.7.1.5 Copper Nanoparticles

The oil, protein, and ash content of *Helianthus annuus* seeds increased when they were successfully treated, and the oleic acid content of the oil increased as well. Although the total copper concentration was increased, no attempts were made to determine whether or not nanoparticles were present in the seeds (Churilov et al., 2021). According to the findings of another study, irrigation with copper nanoparticles (1 g/ha) increased cytokinin, gibberellin, and indole acetic acid levels in *Vicia sativa* L. while decreasing abscisic acid levels, thereby promoting the development of the plant. However, when applied at doses of 0.03 and 0.05 g/ha, copper nanoparticles produce measurable amounts of the drug's beneficial effects on wheat germination, plant survival rate, fresh and dry weight, and yield per hectare of land planted with wheat.

8.7.1.6 Ferric Oxide Nanoparticles

After soil application of 1 g/kg Fe_2O_3 nanoparticles and 45.87 mg/kg EDTA-Fe, the dry matter biomass and chlorophyll content of *Arachis hypogea* were equivalent to each other. The presence of nanoparticles increased the number of oxygen radicles that act as growth promoters to a certain extent, indicating a hormesis effect. Changes in the concentrations of plant hormones also aided in the development of the plant, this could be a benefit of nanoparticles because they can be adsorbed on sandy soil particles, increasing the amount of nutrients available to plants. After 30 minutes at pH1, Fe_2O_3 nanoparticles (spherical and hexagonal, 8 nm but agglomerated 100 mg/litre) released 20% of their iron content (agglomerated). A significant proportion of these Fe^{3+} ions was also found to be tightly bound to the nanoparticle surface, which was another finding (Knijnenburg et al., 2014).

8.7.2 Nanoparticles as Nano-pesticides

8.7.2.1 Silica Nanoparticles

Silica nanoparticles were used to protect *Oryza sativa* grains from weevils at concentrations of 0.5 and 1 g/kg (nanoparticle per grain) (*Sitophilus oryzae*). When compared to finely ground silica, bulk-sized silica killed significantly fewer weevils. Following the treatment of *Oryza sativa* grains with these densities of silica nanoparticles, no new weevil progeny were discovered, indicating that the treatment was successful (Debnath et al., 2011). Silica nanoparticles can emit silicate ions, albeit in trace amounts. However, rather than on the pests, silica ions may have an insecticidal effect on the grain itself. One method is to strengthen the cell wall through the deposition of solid silica, and another is to stimulate the biosynthesis of defence compounds, most likely by acting as a ligand for organic metabolites, both of which are possible (Epstein, 2009).

8.7.2.2 Sulphur Nanoparticles

Due to the rarity of sulphur nanoparticles reducing spontaneously to sulphide ions, the formation of these particles necessitates the use of polysulphides, this may cause a similar amount of residual sulphide ion to be absorbed onto their surface due to the process. The compounds can be dissolved in water, where they are hydrolyzed to produce hydrogen sulphide, a well-known antifungal agent. When applied in the form of a 150 ml NaHS solution (0.01–2.5 mM), which hydrolyzed to form H_2S in a sealed container with a volume of 3 litres, hydrogen sulphide exhibited a dose-dependent microbicidal effect against *Rhizopus oryzae*, *Aspergillus niger*, and *Penicillium italicum* (the effect was demonstrated *in vitro* and on inoculated fruits of *Malus domestica*, *Citrus sine*) (Fu et al., 2014).

8.7.2.3 Gold Nanoparticles

With the help of gold nanoparticles, researchers demonstrated activity against *Aspergillus flavus*, *Aspergillus niger*, and *Puccinia graminis* in the laboratory. *In vitro*, other gold nanomaterials were more promising against *Aspergillus niger*, *Fusarium oxysporum*, and *E. coli*. In vitro tests revealed that gold nanoparticles were effective against *Aspergillus flavus* and *Vibrio cholerae*, respectively. The activity of the nanoparticles was attributed to cell damage (Lokina et al., 2014).

8.7.2.4 Silver Nanoparticles

The antifungal activity of silver nanoparticles could also be a result of the release of silver ions from the particles. Using concentrations ranging from 70 to 800 g/ml, silver nitrate was found to stop the growth of mycotoxigenic *Aspergillus flavus* OC1 strains, as well as the growth of *Penicillium vulpinum* (*Penicillium flavum*) CM1 strains at concentrations ranging from 60 to 800 g/ml (Ismaiel and Tharwat, 2014).

The antibacterial action of silver nanoparticles has been observed in other types of silver nanoparticles. In addition to *Bacillus megaterium* and *Pseudomonas syringae*, it has antibacterial activity against *Xanthomonas oryzae* and *Bacillus thuringiensis*, among other bacteria (Velmurugan et al., 2013). A silver nanoparticle concentration of 0.3–0.75 parts per million demonstrated dose-dependent cytotoxic activity against *Nitrosomonas europaea*, a bacterial infection that degrades ammonium-based fertilizers by oxidizing ammonia to form nitrite, according to the results of the study (Radniecki et al., 2011). It has been discovered that silver ions are toxic to the bacteria *Nitrosomonas europaea* with a toxic range of 0.05–1.25 parts per million (ppm). It has also been found that three various types of silver nanoparticles have a dose-dependent effect on the bacteria *Nitrosomonas europaea* at concentrations ranging from 1–10 mg/litre. During the toxicity test, it was discovered that the bacterial cells were trapped within an aggregate of nanomaterials, which was confirmed by other researchers (Yuan et al., 2013).

8.7.2.5 Platinum Nanoparticles

Platinum nanoparticles demonstrated exceptional activity against two plant pathogenic fungi (*Colletotrichum acutum* and *Cladosporium fulvum*) at concentrations of 2, 4, and 8 g/well. However, they were ineffective against *Phytophthora capsica*, *Phytophthora drechsleri*, or *Didymella bryoniae* when the concentrations were 2, 4, and 8 g/well, respectively (Velmurugan et al., 2016).

8.7.2.6 Chitosan Nanoparticles

In vitro, chitosan nanoparticles demonstrated dose-dependent activity against *Fusarium graminearum* at concentrations ranging from 100 to 5000 parts per million (ppm); at 500 and 1000 parts per million (ppm), they also provided protection to infected wheat spikelets. Chitosan in bulk had the same antifungal activity as chitosan in suspension. Aggregated nanomaterials were discovered inside the spikelet and leaf cells of plants that had been sprayed with nanoparticles, and their size orientation did not match that of the nanoparticles that had been applied (Alghuthaymi et al., 2021). Although the inhibitory effect of chitosan nanoparticles and microparticles was not statistically significant, it was found to be significantly greater than that of bulk chitosan at a concentration of 5.10^3 g per petri dish on *Aspergillus parasiticus* spore germination and growth *in vitro*. The amount of chitosan and sodium tripolyphosphate used in the manufacturing process, as well as the availability of free phosphate groups that chelated $Ca^{2+\cdot}$ Fe^{2+}, and Mg^{2+} ions and made them available to fungal cells, were found to be important factors in the antifungal activity (Cota-Arriola et al., 2013). Chitosan nanoparticle-based ion gelation method is used extensively in plant disease management systems. Many studies proposed the use of chitosan-based nanoparticles as nanopesticides, nanoherbicides and nanofertilizers (Bocchetta, 2020; Hoang et al., 2022; Malerba and Cerana, 2019; Maluin and Hussein, 2020; Rodriguez et al., 2019) (Table 8.1).

TABLE 8.1
Types of Nanoparticles Used as Antimicrobial Agents or Carriers of Antimicrobials

Nanostructured material	Characteristics	Growth plant stimulating effect	Pesticidal effect	Pesticide carried	Phytotoxicity	References
SiO_2 (Silicon dioxide)	Lipophilic and hydrophilic, average particle size 12 nm , surface area 200 m^2/g Functionalized with 3-aminopropyl groups, 80-100 nm diameter, shell thickness 15 nm Spherical average diameter 320 nm, pore diameter 12 nm, coated with polyvinylpyrrolidone Outside diameter 15–20 nm Spherical, uncoated, hydrodynamic diameter 135.2 ± 1.1 and 291.6 ± 2.3 nm, zeta potential -20.3 ± 0.4 X-ray diffraction crystalline structure, FTIR- 4000-400 cm^{-1}	*Lycopersicon esculentum* *Lactuca sativa* *Oryza sativa* (with encapsulated uniconazole) Promotion of germination, growth and production of *Eruca sativa*, antifungal against *Fusarium oxysporum* and *Aspergillus niger*	- - *Sitophillous oryzae* on rice	- - Albamectine	*Arabidopsis thaliana*	(Debnath et al., 2011; Shah and Belozerova, 2009; Siddiqui and Al-Whaibi, 2014; Slomberg and Schoenfisch, 2012; Tan et al., 2012; Wang et al., 2014) (Goswami and Mathur, 2022)

(*continued*)

TABLE 8.1 (Continued)
Types of Nanoparticles Used as Antimicrobial Agents or Carriers of Antimicrobials

Nanostructured material	Characteristics	Growth plant stimulating effect	Pesticidal effect	Pesticide carried	Phytotoxicity	References
S (Sulphur)	Stabilized with PEG-400, hydrodynamic diameter 20–70 nm and 10–100 nm, particle size 15–20 nm, mean surface roughness 5.313 and 3.840 nm respectively Orthorhombic: spherical, contain 38.68 wt.% of S; monoclinic: cylindrical, stabilized with conjugate of Span 80 and tween	*Aspergillus niger*, *fusarium oxysporum*	*Aspergillus niger*			(Choudhury et al., 2012, 2011)
Se (Selenium)	-	*Zea mays*				(Gudkov et al., 2020)
Au (Gold)	Dodecanethiol functionalized Synthesized by the citrate method, spherical, 10–20 nm	*Lactuca sativa* *Brassica juncea*				(Khanna and Zaidi, 2012)
Ag (Silver)	Alginate-capped Citrate capped, 6–42 nm. Average size 29 nm Glycoxylan-coated, average size 8, 10 or 25 nm Spherical, 70–140 nm, prepared using *Manilkara zapota* extract	*Triticum aestivum* *Brassica juncea* *Raphanus sativus* L.	*Rhipicephalus (Boophilus) microplus* larvae			(Yurkova et al., 2020) (Sah et al., 2011) (Iram et al., 2014) (Rajakumar and Rahuman, 2012)
$\mathbf{Ag_2O/Ag}$ (Silver oxide)	Spherical, 34±2 nm, contain small amount of metallic silver		*Aspergillus niger*, *Aspergillus falvus* *Aspergillus funigatus*			(Thenmozhi et al., 2013)
$\mathbf{Ag3O_4}$ (Silver oxide)	18.72 nm		*Aspergillus niger*			(Azam et al., 2012) (Suresh et al., 2016)
Cu (Copper)	-	*Lactuca sativa*				(Shah and Belozerova, 2009)
CuO (Copper oxide)	25 nm, surface area up to 25 m^2/g	*Triticum aestivum*				(Churilov et al., 2021)
Pt (Platinum)	Spherical and oval, size 10–50 nm, synthesized with Yoshino cherry tree		*Cladosporium fulvum*, *Colletotrichum acutatum*			(Velmurugan et al., 2016)
ZnO (Zinc oxide)	17 nm, +20.7mV surface charge 17.4 ± 4.9 nm, 11.0 ± 0.7 nm hydrodynamic diameter			Thiram	*Brassica oleracea*, *Zea mays*	(Xue et al., 2014) (Pokhrel and Dubey, 2013)

8.8 FUTURE PERSPECTIVE

Given the chemical makeup of nanomaterials, it is critical to evaluate the possibility of negative consequences, such as phytotoxicity at high toxicity and concentrations to unintended non-target organisms. Sadly, there is currently less information available on the ultimate fate of nanomaterials, particularly those that did not reach their intended target and escaped from an agricultural area, as of this writing. Notably, even nanoparticles manufactured using "green" methods are not always 100% "green", as demonstrated. The toxicity of nanoparticles to humans must also be considered, as nanoparticles are frequently unable to be eliminated from agricultural products that humans consume. Before any form of nanoparticle can be used in practical agricultural applications, it is necessary to evaluate all potential negative effects. The vast majority of papers contain only preliminary findings that do not take into consideration any of these issues.

The difficulty of the synthetic methods under consideration to be scaled up is another concern that may develop. For a more practical plan to be developed, future research must be carried out in-depth to examine the level of risk connected with NP-based products in agricultural production.

The second goal is to get a comprehensive understanding of NPs bioaccumulation capacity during environmental clean-up and its impact on nanotoxicity and develop our understanding of tropical chain transmission within the ecosystem chain and decode the sufficient safety estimate. Food nanotechnology and its applications should be made available to the public through a government-sponsored programme, including the development of a database that would benefit both the general public and agricultural processed food manufacturers. Determining the lowest possible concentration of NP in natural soil should be a priority to create a nontoxic natural soil system that may be employed in field applications.

The use of nanomaterials such as nanofertilizers, nanopesticides, and nanoherbicides has significantly increased crop yields worldwide. However, more research is essential to determine their method of action and any potential hazards they may pose after entering the food chain.

In field-based research, the NP-rhizome-microbiome relationship should be thoroughly investigated. Biogenic nanoparticle production can be achieved using biocontrol fungi such as *Trichoderma*. One of the main areas of future research will be the development of innovative and safe technologies for biogenic synthesis of nanoparticles without the use of hazardous substances.

Multidrug resistance is an unavoidable problem due to widespread antibiotic usage. However, a thorough understanding of the many mechanisms of resistance in phytopathogens is a critical first step in combating future antibiotic resistance. Similarly, finding new targets for nanoparticle-microbe interactions is a top priority in the fight against emerging diseases.

ACKNOWLEDGMENT

The authors are thankful to Dr. Ramdas Atram (Director, Govt. Institute of Forensic Science, Nagpur) for constant encouragement and support.

REFERENCES

Adisa, I.O., Pullagurala, V.L.R., Peralta-Videa, J.R., Dimkpa, C.O., Elmer, W.H., Gardea-Torresdey, J.L., White, J.C., 2019. Recent advances in nano-enabled fertilizers and pesticides: a critical review of mechanisms of action. *Environ. Sci.: Nano.* 6, 2002–2030. https://doi.org/10.1039/C9EN00265K

Agrios, G.N., 2009. Plant Pathogens and Disease: General Introduction, in: Schaechter, M. (Ed.), *Encyclopedia of Microbiology* (Third Edition). Academic Press, Oxford, pp. 613–646. https://doi.org/10.1016/B978-012373944-5.00344-8

Alghuthaymi, M.A., Almoammar, H., Rai, M., Said-Galiev, E., Abd-Elsalam, K.A., 2015. Myconanoparticles: synthesis and their role in phytopathogens management. *Biotechnol. Biotechnol. Equip.* 29, 221–236. https://doi.org/10.1080/13102818.2015.1008194

Alghuthaymi, M.A., Hassan, A.A., Kalia, A., Sayed El Ahl, R.M., El Hamaky, A.A., Oleksak, P., Kuca, K., Abd-Elsalam, K.A., 2021. Antifungal Nano-Therapy in Veterinary Medicine: Current Status and Future Prospects. *J. Fungi.* 7, 494.

Alsuwayyid, A.A., Alslimah, A.S., Perveen, K., Bukhari, N.A., Al-Humaid, L.A., 2022. Effect of zinc oxide nanoparticles on Triticum aestivum L. and bioaccumulation assessment using ICP-MS and SEM analysis. *J. King Saud Univ. Sci.* 34, 101944. https://doi.org/10.1016/j.jksus.2022.101944

Alvarez-Puebla, R.A., Jr, D.S.D.S., Aroca, R.F., 2004. Surface-enhanced Raman scattering for ultrasensitive chemical analysis of 1 and 2-naphthalenethiols. *Analyst* 129, 1251–1256. https://doi.org/10.1039/B410488A

Amendola, V., Polizzi, S., Meneghetti, M., 2006. Laser ablation synthesis of gold nanoparticles in organic solvents. *J. Phys. Chem. B.* 110(14), 7232–7237.

Anandhi, S., Saminathan, V., Yasotha, P., Saravanan, P., Rajanbabu, V., 2020. Nano-pesticides in pest management. *J Entomol Zool Stud.* 8, 685–90.

Anderson, H.W., Gottlieb, D., 1952. Plant disease control with antibiotics. *Econ Bot.* 6, 294–308. https://doi.org/10.1007/BF02985071

Anjum, M., Oves, M., Kumar, R., Barakat, M., 2017. Fabrication of ZnO-ZnS@ polyaniline nanohybrid for enhanced photocatalytic degradation of 2-chlorophenol and microbial contaminants in wastewater. *International Biodeterioration & Biodegradation* 119, 66–77.

Aslam, M., Gopakumar, G., Shoba, T.L., Mulla, I.S., Vijayamohanan, K., Kulkarni, S.K., Urban, J., Vogel, W., 2002. Formation of Cu and Cu2O nanoparticles by variation of the surface ligand: Preparation, structure, and insulating-to-metallic transition. *J Colloid Interface Sci.* 255, 79–90. https://doi.org/10.1006/jcis.2002.8558

Azam, A., Ahmed, A.S., Oves, M., Khan, M., Memic, A., 2012. Size-dependent antimicrobial properties of CuO nanoparticles against Gram-positive and-negative bacterial strains. *Int J Nanomedicine* 7, 3527.

Aziz, N., Fatma, T., Varma, A., Prasad, R. 2014. Biogenic synthesis of silver nanoparticles using *Scenedesmus abundans* and evaluation of their antibacterial activity. Journal of Nanoparticles. Article ID 689419, 2014. http://dx.doi. org/10.1155/2014/689419

Aziz, N., Faraz, M., Pandey, R., Sakir, M., Fatma, T., Varma, A., Barman, I., Prasad, R. 2015. Facile algae-derived route to biogenic silver nanoparticles: Synthesis, antibacterial and photocatalytic properties. *Langmuir.* 31, 11605–11612. doi 10.1021/acs.langmuir.5b03081

Aziz, N., Pandey, R., Barman, I., Prasad, R. 2016. Leveraging the attributes of *Mucor hiemalis*-derived silver nanoparticles for a synergistic broad-spectrum antimicrobial platform. Front Microbiol. 7(1984). doi: 10.3389/fmicb.2016.01984

Aziz, N., Faraz, M., Sherwani, M. A., Fatma, T., Prasad, R. 2019. Illuminating the anticancerous efficacy of a new fungal chassis for silver nanoparticle synthesis. *Front Chem.* 7(65). doi: 10.3389/fchem.2019.00065

Bahrulolum, H., Nooraei, S., Javanshir, N., Tarrahimofrad, H., Mirbagheri, V.S., Easton, A.J., Ahmadian, G., 2021. Green synthesis of metal nanoparticles using microorganisms and their application in the agrifood sector. *J. Nanobiotechnology* 19, 86. https://doi.org/10.1186/s12951-021-00834-3

Bano, I., Skalickova, S., Sajjad, H., Skladanka, J., Horky, P., 2021. Uses of Selenium Nanoparticles in the Plant Production. *Agron.* 11, 2229. https://doi.org/10.3390/agronomy11112229

Bhat, J.A., Rajora, N., Raturi, G., Sharma, S., Dhiman, P., Sanand, S., Shivaraj, S.M., Sonah, H., Deshmukh, R., 2021. Silicon nanoparticles (SiNPs) in sustainable agriculture: major emphasis on the practicality, efficacy and concerns. *Nanoscale Adv.* 3, 4019–4028. https://doi.org/10.1039/D1NA00233C

Blosi, M., Albonetti, S., Dondi, M., Martelli, C., Baldi, G., 2011. Microwave-assisted polyol synthesis of Cu nanoparticles. *J Nanopart Res.* 13, 127–138. https://doi.org/10.1007/s11051-010-0010-7

Bocchetta, P., 2020. Ionotropic gelation of chitosan for next-generation composite proton conducting flat structures. *Molecules* 25, 1632.

Byrappa, K., Yoshimura, M., 2012. *Handbook of hydrothermal technology*. William Andrew.

Cai, L., Chen, J., Liu, Z., Wang, H., Yang, H., Ding, W., 2018. Magnesium Oxide Nanoparticles: Effective Agricultural Antibacterial Agent Against Ralstonia solanacearum. *Front. Microbiol.* 9, 790. https://doi.org/10.3389/fmicb.2018.00790

Carter, A.P., Clemons, W.M., Brodersen, D.E., Morgan-Warren, R.J., Wimberly, B.T., Ramakrishnan, V., 2000. Functional insights from the structure of the 30S ribosomal subunit and its interactions with antibiotics. *Nature* 407, 340–348.

Chaudhry, N., Dwivedi, S., Chaudhry, V., Singh, A., Saquib, Q., Azam, A., Musarrat, J., 2018. Bio-inspired nanomaterials in agriculture and food: Current status, foreseen applications and challenges. *Microb. Pathog.* 123, 196–200. https://doi.org/10.1016/j.micpath.2018.07.013

Chen, L., Zhang, D., Chen, J., Zhou, H., Wan, H., 2006. The use of CTAB to control the size of copper nanoparticles and the concentration of alkylthiols on their surfaces. *Mater. Sci. Eng.: A* 415, 156–161. https://doi.org/10.1016/j.msea.2005.09.060

Choi, O., Yu, C.-P., Esteban Fernández, G., Hu, Z., 2010. Interactions of nanosilver with Escherichia coli cells in planktonic and biofilm cultures. *Water Res.* 44, 6095–6103. https://doi.org/10.1016/j.watres.2010.06.069

Chongdar, S., Gunasekaran, G., Kumar, P., 2005. Corrosion inhibition of mild steel by aerobic biofilm. *Electrochimi. Acta* 50, 4655–4665. https://doi.org/10.1016/j.electacta.2005.02.017

Chopra, I., Roberts, M., 2001. Tetracycline antibiotics: Mode of action, applications, molecular biology, and epidemiology of bacterial resistance. *Microbiol Mol Biol Rev.* 65, 232–260. https://doi.org/10.1128/MMBR.65.2.232-260.2001

Choudhury, S.R., Ghosh, M., Mandal, A., Chakravorty, D., Pal, M., Pradhan, S., Goswami, A., 2011. Surface-modified sulfur nanoparticles: an effective antifungal agent against Aspergillus niger and Fusarium oxysporum. *Appl. Microbiol. Biotechnol.* 90, 733–743.

Choudhury, S.R., Ghosh, M., Goswami, A., 2012. Inhibitory effects of sulfur nanoparticles on membrane lipids of Aspergillus niger: A novel route of fungistasis. *Curr. Microbiol.* 65, 91–97.

Churilov, D., Churilov, G., Churilova, V., Polischuk, S., Borychev, S., Byshov, N., 2021. Influence of nanoparticles on soil microflora. *Int. J. Nanotechnol.* 18, 788–802.

Clark, R.B., Hunt, D.K., He, M., Achorn, C., Chen, C.-L., Deng, Y., Fyfe, C., Grossman, T.H., Hogan, P.C., O'Brien, W.J., 2012. Fluorocyclines. 2. Optimization of the C-9 side-chain for antibacterial activity and oral efficacy. *J. Med. Chem.* 55, 606–622.

Cota-Arriola, O., Cortez-Rocha, M.O., Ezquerra-Brauer, J.M., Lizardi-Mendoza, J., Burgos-Hernández, A., Robles-Sánchez, R.M., Plascencia-Jatomea, M., 2013. Ultrastructural, morphological, and antifungal properties of micro and nanoparticles of chitosan crosslinked with sodium tripolyphosphate. *J Polym Environ.* 21, 971–980.

Cunha, B.A., Sibley, C.M., Ristuccia, A.M., 1982. Doxycycline. *Ther Drug Monit* 4, 115–135.

De Jong, W.H., Borm, P.J., 2008. Drug delivery and nanoparticles: Applications and hazards. *Int J Nanomedicine* 3, 133.

Debnath, N., Das, S., Seth, D., Chandra, R., Bhattacharya, S.C., Goswami, A., 2011. Entomotoxic effect of silica nanoparticles against Sitophilus oryzae (L.). *J. Pest Sci.* 84, 99–105.

Dhiman, S., Varma, A., Prasad, R., Goel, A. 2022. Mechanistic insight of the antifungal potential of green synthesized zinc oxide nanoparticles against *Alternaria brassicae. Journal of Nanomaterials.* 13 pages. https://doi.org/10.1155/2022/7138843

Dhiman, S., Varma, A., Rao, M., Prasad, R., Goel, A. 2023. Deciphering the fertilizing and disease suppression potential of phytofabricated zinc oxide nanoparticles on *Brassica juncea. Environmental Research.* https://doi.org/10.1016/j.envres.2023.116276

Doyle, D., McDowall, K., Butler, M., Hunter, I., 1991. Characterization of an oxytetracycline-resistance gene, otrA, of Streptomyces rimosus. *Mol. Microbiol.* 5, 2923–2933.

Epstein, E., 2009. Silicon: Its manifold roles in plants. *Ann. Appl. Biol.* 155, 155–160.

Erickson, B., 2021. Groups challenge approval of streptomycin use on citrus. *Chem. Eng. News* 2021, 14.

Fang, J., Zhong, C., Mu, R., 2005. The study of deposited silver particulate films by simple method for efficient SERS. *Chem. Phys. Lett.* 401, 271–275.

Fu, L.-H., Hu, K.-D., Hu, L.-Y., Li, Y.-H., Hu, L.-B., Yan, H., Liu, Y.-S., Zhang, H., 2014. An antifungal role of hydrogen sulfide on the postharvest pathogens Aspergillus niger and Penicillium italicum. *PLoS One* 9, e104206.

Gahlawat, G., Choudhury, A.R., 2019. A review on the biosynthesis of metal and metal salt nanoparticles by microbes. *RSC Adv.* 9, 12944–12967. https://doi.org/10.1039/C8RA10483B

Gallo, V., Zappettini, A., Villani, M., Marmiroli, N., Marmiroli, M., 2021. Comparative Analysis of Proteins Regulated during Cadmium Sulfide Quantum Dots Response in Arabidopsis thaliana Wild Type and Tolerant Mutants. *Nanomater.* 11, 615.

Garza-García, J.J.O., Hernández-Díaz, J.A., Zamudio-Ojeda, A., León-Morales, J.M., Guerrero-Guzmán, A., Sánchez-Chiprés, D.R., López-Velázquez, J.C., García-Morales, S., 2022. The Role of Selenium Nanoparticles in Agriculture and Food Technology. *Biol Trace Elem Res.* 200, 2528–2548. https://doi.org/10.1007/s12011-021-02847-3

Goswami, P., Mathur, J., 2022. Application of agro-waste-mediated silica nanoparticles to sustainable agriculture. *Bioresources and Bioprocessing* 9, 9. https://doi.org/10.1186/s40643-022-00496-5

Gudikandula, K., Charya Maringanti, S., 2016. Synthesis of silver nanoparticles by chemical and biological methods and their antimicrobial properties. *J. Exp. Nanosci.* 11, 714–721. https://doi.org/10.1080/17458080.2016.1139196

Gudkov, S.V., Shafeev, G.A., Glinushkin, A.P., Shkirin, A.V., Barmina, E.V., Rakov, I.I., Simakin, A.V., Kislov, A.V., Astashev, M.E., Vodeneev, V.A., 2020. Production and use of selenium nanoparticles as fertilizers. *ACS Omega* 5, 17767–17774.

Guillaume, G., Ledent, V., Moens, W., Collard, J.-M., 2004. Phylogeny of efflux-mediated tetracycline resistance genes and related proteins revisited. *Microbial Drug Resistance* 10, 11–26.

Gunasekaran, G., Chongdar, S., Gaonkar, S.N., Kumar, P., 2004. Influence of bacteria on film formation inhibiting corrosion. *Corros. Sci.* 46, 1953–1967. https://doi.org/10.1016/j.corsci.2003.10.023

Hajipour, M.J., Fromm, K.M., Akbar Ashkarran, A., Jimenez de Aberasturi, D., Larramendi, I.R. de, Rojo, T., Serpooshan, V., Parak, W.J., Mahmoudi, M., 2012. Antibacterial properties of nanoparticles. *Trends Biotechnol.* 30, 499–511. https://doi.org/10.1016/j.tibtech.2012.06.004

Hallmann, K.-C.I.M.K., Macierzanka, E.K.E.P.T., 2017. A. et al., Rhamnolipid CMC prediction. *J. Colloid Interface Sci.* 488, 10–19.

Han, Y., Luo, Y., Qin, S., Xi, L., Wan, B., Du, L., 2014. Induction of systemic resistance against tobacco mosaic virus by Ningnanmycin in tobacco. *Pestic. Biochem. Phys.* 111, 14–18.

Hoang, N.H., Le Thanh, T., Sangpueak, R., Treekoon, J., Saengchan, C., Thepbandit, W., Papathoti, N.K., Kamkaew, A., Buensanteai, N., 2022. Chitosan Nanoparticles-Based Ionic Gelation Method: A Promising Candidate for Plant Disease Management. *Polymers (Basel)* 14, 662. https://doi.org/10.3390/polym14040662

Hollingshead, S., Vapnek, D., 1985. Nucleotide sequence analysis of a gene encoding a streptomycin/spectinomycin adenylyltransferase. *Plasmid* 13, 17–30. https://doi.org/10.1016/0147-619x(85)90052-6

Husen, A., Jawaid, M., 2020. *Nanomaterials for agriculture and forestry applications*. Elsevier.

Hwang, M.S., Morgan, R.L., Sarkar, S.F., Wang, P.W., Guttman, D.S., 2005. Phylogenetic characterization of virulence and resistance phenotypes of Pseudomonas syringae. *Appl Environ Microbiol* 71, 5182–5191.

Ibrahim, E., Zhang, M., Zhang, Y., Hossain, A., Qiu, W., Chen, Y., Wang, Y., Wu, W., Sun, G., Li, B., 2020. Green-synthesization of silver nanoparticles using endophytic bacteria isolated from garlic and its antifungal activity against wheat fusarium head blight pathogen fusarium graminearum. *Nanomater.* 10. https://doi.org/10.3390/nano10020219

Iram, F., Iqbal, M.S., Athar, M.M., Saeed, M.Z., Yasmeen, A., Ahmad, R., 2014. Glucoxylan-mediated green synthesis of gold and silver nanoparticles and their phyto-toxicity study. *Carbohydr. Polym.* 104, 29–33.

Ismaiel, A., Tharwat, N., 2014. Antifungal activity of silver ion on ultrastructure and production of aflatoxin B1 and patulin by two mycotoxigenic strains, Aspergillus flavus OC1 and Penicillium vulpinum CM1. *Journal de mycologie médicale* 24, 193–204.

Jaiswal, S., Duffy, B., Jaiswal, A.K., Stobie, N., McHale, P., 2010. Enhancement of the antibacterial properties of silver nanoparticles using β-cyclodextrin as a capping agent. *Int. J. Antimicrob. Agents.* 36, 280–283. https://doi.org/10.1016/j.ijantimicag.2010.05.006

Jarolmen, H., Hewel, D., Kain, E., 1970. Activity of minocycline against R factor-carrying Enterobacteriaceae. *Infect. Immu.* 1, 321–326.

Jośko, I., Oleszczuk, P., Skwarek, E., 2017. Toxicity of combined mixtures of nanoparticles to plants. *J. Hazard. Mater.* 331, 200–209. https://doi.org/10.1016/j.jhazmat.2017.02.028

Khan, M., Ahamad, F., Fatima, T., 2019. Effect of nanoparticles on plant pathogens. pp. 215–240. https://doi.org/10.1016/B978-0-12-815322-2.00009-2

Khan, M.R., Rizvi, T.F., 2014. Nanotechnology: Scope and application in plant disease management. *Plant Pathol J.* 13, 214–231.

Khanna, P., Zaidi, M., 2012. Sandeep Arora, Priyadarshini Sharma, Sumit Kumar, Rajeev Nayan. *Plant Growth Regul.* 66, 303–310.

Kisimba, K., Krishnan, A., Faya, M., Byanga, K., Kasumbwe, K., Vijayakumar, K., Prasad, R. 2023. Synthesis of metallic nanoparticles based on green chemistry and their medical biochemical applications: Synthesis of metallic nanoparticles. *Journal of Renewable Materials.* 11(6), 2575–2591. https://doi.org/10.32604/jrm.2023.026159

Kitchens, C.L., McLeod, M.C., Roberts, C.B., 2005. Chloride Ion effects on synthesis and directed assembly of copper nanoparticles in liquid and compressed alkane microemulsions. *Langmuir* 21, 5166–5173. https://doi.org/10.1021/la047785x

Knijnenburg, J.T., Seristatidou, E., Hilty, F.M., Krumeich, F., Deligiannakis, Y., 2014. Proton-promoted iron dissolution from nanoparticles and the influence by the local iron environment. *J. Phys. Chem. C.* 118, 24072–24080.

Komarneni, S., 2003. Nanophase materials by hydrothermal, microwave-hydrothermal and microwave-solvothermal methods. *Curr. Sci.* 85, 1730–1734.

Kumar, Vineet, Guleria, P., Kumar, Vinay, Yadav, S.K., 2013. Gold nanoparticle exposure induces growth and yield enhancement in Arabidopsis thaliana. *Sci. Total Environ.* 461, 462–468.

Kuntyi, O., Mazur, A., Kytsya, A., Karpenko, O., Bazylyak, L., Mertsalo, I., Pokynbroda, T., Prokopalo, A., 2020. Electrochemical synthesis of silver nanoparticles in solutions of rhamnolipid. *Micro. Nano. Lett.* 15, 802–807. https://doi.org/10.1049/mnl.2020.0195

Leben, Curt., Keitt, G.W., 1954. Antibiotics and plant disease, effects of antibiotics in control of plant diseases. *J. Agric. Food Chem.* 2, 234–239. https://doi.org/10.1021/jf60025a003

Lee, H., Kim, Y., Kim, J., Choi, G., Park, S., Kim, C., Jung, H., 2005. Activity of some aminoglycoside antibiotics against true fungi, Phytophthora and Pythium species. *J. Appl. Microbiol.* 99, 836–843.

Lemire, J.A., Harrison, J.J., Turner, R.J., 2013. Antimicrobial activity of metals: Mechanisms, molecular targets and applications. *Nat Rev Microbiol.* 11, 371–384. https://doi.org/10.1038/nrmicro3028

Liu, X., Wang, J., Zhang, J., Liu, B., Zhou, J., Yang, S., 2007. Fabrication and characterization of Ag/polymer nanocomposite films through layer-by-layer self-assembly technique. *Thin Solid Films* 515, 7870–7875. https://doi.org/10.1016/j.tsf.2007.03.054

Lokina, S., Suresh, R., Giribabu, K., Stephen, A., Sundaram, R.L., Narayanan, V., 2014. Spectroscopic investigations, antimicrobial, and cytotoxic activity of green synthesized gold nanoparticles. *Spectrochim Acta A Mol Biomol Spectrosc* 129, 484–490.

Loper, J., Henkels, M., Roberts, R., Grove, G., Willet, M., 1991. Evaluation of streptomycin, oxytetracycline, and copper resistance of Erwinia amylovora isolated from pear orchards in Washington State. *Plant Dis.* 75, 287–290.

Luo, Z.-Q., Farrand, S.K., 1999. Cloning and characterization of a tetracycline resistance determinant present in Agrobacterium tumefaciens C58. *J. Bacteriol.* 181, 618–626.

Ma, J., Nishimura, K., Takahashi, E., 1989. Effect of silicon on the growth of rice plant at different growth stages. *J. Soil Sci. Plant Nutr.* 35, 347–356.

Malerba, M., Cerana, R., 2019. Recent applications of chitin-and chitosan-based polymers in plants. *Polymers* 11, 839.

Maluin, F.N., Hussein, M.Z., 2020. Chitosan-based agronanochemicals as a sustainable alternative in crop protection. *Molecules* 25, 1611.

Masum, M.M.I., Siddiqa, M.M., Ali, K.A., Zhang, Y., Abdallah, Y., Ibrahim, E., Qiu, W., Yan, C., Li, B., 2019. Biogenic synthesis of silver nanoparticles using phyllanthus emblica fruit extract and its inhibitory action against the pathogen acidovorax oryzae strain RS-2 of rice bacterial brown stripe. *Front Microbiol.* 10, 820. https://doi.org/10.3389/fmicb.2019.00820

Mazodier, P., Cossart, P., Giraud, E., Gasser, F., 1985. Completion of the nucleotide sequence of the central region of Tn5 confirms the presence of three resistance genes. *Nucleic Acids Res.* 13, 195–205. https://doi.org/10.1093/nar/13.1.195

McManus, P., Jones, A., 1994. *Epidemiology and genetic analysis of streptomycin-resistant Erwinia amylovora from Michigan and evaluation of oxytetracycline for control*. Phytopathology (USA).

McManus, P., Stockwell, V., Sundin, G., Jones, A., 2002. Antibiotic use in plant agriculture. *Annu. Rev. Phytopathol.* 40, 443–65. https://doi.org/10.1146/annurev.phyto.40.120301.093927

Meiliefiana, M., Nakayashiki, T., Yamamoto, E., Hayashi, K., Ohtani, M., Kobiro, K., 2022. One-Step Solvothermal Synthesis of Ni Nanoparticle Catalysts Embedded in ZrO2 Porous Spheres to Suppress Carbon Deposition in Low-Temperature Dry Reforming of Methane. *Nanoscale Res. Lett.* 17, 47. https://doi.org/10.1186/s11671-022-03683-7

Mishra, S., Singh, B.R., Naqvi, A.H., Singh, H.B., 2017. Potential of biosynthesized silver nanoparticles using Stenotrophomonas sp. BHU-S7 (MTCC 5978) for management of soil-borne and foliar phytopathogens. *Sci Rep.* 7, 45154. https://doi.org/10.1038/srep45154

Mogharabi, M., Abdollahi, M., Faramarzi, M.A., 2014. Toxicity of nanomaterials; an undermined issue. *DARU J. Pharm. Sci.* 22, 1–4.

Mohanpuria, P., Rana, N.K., Yadav, S.K., 2008. Biosynthesis of nanoparticles: Technological concepts and future applications. *J Nanopart Res.* 10, 507–517.

Moller, W.J., Schroth, M.N., Thomson, S.V., 1981. The scenario of fire blight and streptomycin resistance. *Plant Dis.* 65,563–568. 10.1094/PD-65-563

Nair, R., 2016. Effects of nanoparticles on plant growth and development, in: Kole, C., Kumar, D.S., Khodakovskaya, M.V. (Eds.), *Plant Nanotechnology: Principles and Practices*. Springer International Publishing, Cham, pp. 95–118. https://doi.org/10.1007/978-3-319-42154-4_5

Narayanan, K.B., Sakthivel, N., 2010. Biological synthesis of metal nanoparticles by microbes. *Adv Colloid Interface Sci.* 156, 1–13. https://doi.org/10.1016/j.cis.2010.02.001

Nelson, M.L., Levy, S.B., 2011. The history of the tetracyclines. *Ann. N. Y. Acad. Sci.* 1241, 17–32.

Nguyen, F., Starosta, A.L., Arenz, S., Sohmen, D., Dönhöfer, A., Wilson, D.N., 2014. Tetracycline antibiotics and resistance mechanisms. *Biol. Chem.* 395, 559–575.

Nixon, W.C., Huxley, H.E., Klug, A., 1971. The general principles of scanning electron microscopy. *Philos. Trans. R. Soc. Lond., B, Biol. Sci.* 261, 45–50. https://doi.org/10.1098/rstb.1971.0035

Perreten, V., Schwarz, F., Cresta, L., Boeglin, M., Dasen, G., Teuber, M., 1997. Antibiotic resistance spread in food. *Nature* 389, 801–802. https://doi.org/10.1038/39767

Pokhrel, L.R., Dubey, B., 2013. Evaluation of developmental responses of two crop plants exposed to silver and zinc oxide nanoparticles. *Sci. Total Environ.* 452–453, 321–332. https://doi.org/10.1016/j.scitotenv.2013.02.059

Prasad, R. (2014). Synthesis of silver nanoparticles in photosynthetic plants. *Journal of Nanoparticles*. Article ID 963961, 2014. http://dx.doi.org/10.1155/2014/963961

Prasad, P.V., Shrivastav, T.G., 2014. Nanotechnological contribution to drug delivery system: A reappraisal. *J. Biomater. Nanobiotechnol.* 2014.

Prasad, R., Kumar, V., Prasad, K.S. (2014). Nanotechnology in sustainable agriculture: present concerns and future aspects. *Afr J Biotechnol.* 13(6), 705–713.

Prasad, R., Pandey, R., Barman, I. (2016). Engineering tailored nanoparticles with microbes: quo vadis. *WIREs Nanomed Nanobiotechnol 8*, 316–330. doi: 10.1002/wnan.1363

Prasad, R., Bhattacharyya, A., Nguyen, Q.D. (2017). Nanotechnology in sustainable agriculture: Recent developments, challenges, and perspectives. *Front Microbiol.* 8, 1014. doi: 10.3389/fmicb.2017.01014

Projan, S.J., Moghazeh, S., Novick, R.P., 1988. Nucleotide sequence of pS194, a streptomycin-resistance plasmid from Staphylococcus aureus. *Nucleic Acids Res.* 16, 2179–2188.

Radniecki, T.S., Stankus, D.P., Neigh, A., Nason, J.A., Semprini, L., 2011. Influence of liberated silver from silver nanoparticles on nitrification inhibition of Nitrosomonas europaea. *Chemosphere* 85, 43–49.

Raja, M., Subha, J., Ali, F.B., Ryu, S.H., 2008. Synthesis of copper nanoparticles by electroreduction process. *Mater. Manuf. Process.* 23, 782–785.

Rajakumar, G., Rahuman, A.A., 2012. Acaricidal activity of aqueous extract and synthesized silver nanoparticles from Manilkara zapota against Rhipicephalus (Boophilus) microplus. *Res. Vet. Sci.* 93, 303–309.

Rajput, V.D., Minkina, T., Feizi, M., Kumari, A., Khan, M., Mandzhieva, S., Sushkova, S., El-Ramady, H., Verma, K.K., Singh, A., Hullebusch, E.D. van, Singh, R.K., Jatav, H.S., Choudhary, R., 2021. Effects of Silicon and Silicon-Based Nanoparticles on Rhizosphere Microbiome, Plant Stress and Growth. *Biology (Basel)* 10, 791. https://doi.org/10.3390/biology10080791

Raman, N.M., Easwaran, M., Kaul, R., Bharti, J., Motelb, K.F.A., Kaul, T., 2020. Antimicrobial resistance with special emphasis on pathogens in agriculture, antimicrobial resistance – A one health perspective. *Intech Open*. https://doi.org/10.5772/intechopen.88440

Revell, P., 2006. The biological effects of nanoparticles. *Nanotec Percep.* 2, 283–298.

Rodriguez, V.A., Bolla, P.K., Kalhapure, R.S., Boddu, S.H.S., Neupane, R., Franco, J., Renukuntla, J., 2019. Preparation and characterization of furosemide-silver complex loaded chitosan nanoparticles. *Processes* 7, 206.

Sah, S., Sorooshzadeh, A., Rezazadehs, H., Naghdibadi, H., 2011. Effect of nano silver and silver nitrate on seed yield of borage. *J. Med. Plant Res.* 5, 171–175.

Sarlak, N., Abdi, M.R., 2022. Chapter 21 – Biosynthesis and green synthesis of nanoparticles and their use in agriculture, in: Ghorbanpour, M., Shahid, M.A. (Eds.), *Nano-Enabled Agrochemicals in Agriculture*. Academic Press, pp. 377–391. https://doi.org/10.1016/B978-0-323-91009-5.00011-2

Seabra, A.B., Rai, M., Durán, N., 2014. Nano carriers for nitric oxide delivery and its potential applications in plant physiological process: A mini review. *J. Plant Biochem. Biotechnol.* 23, 1–10. https://doi.org/10.1007/s13562-013-0204-z

Selyutina, O.Y., Apanasenko, I., Khalikov, S., Polyakov, N.E., 2017. Natural poly-and oligosaccharides as novel delivery systems for plant protection compounds. *J. Agric. Food Chem.* 65, 6582–6587.

Selyutina, O.Y., Khalikov, S.S., Polyakov, N.E., 2020. Arabinogalactan and glycyrrhizin based nanopesticides as novel delivery systems for plant protection. *Environ. Sci. Pollut. Res.* 27, 5864–5872.

Shah, V., Belozerova, I., 2009. Influence of metal nanoparticles on the soil microbial community and germination of lettuce seeds. *Water, Air, and Soil Pollut.* 197, 143–148.

Sharma, P., Bhatt, D., Zaidi, M., Saradhi, P.P., Khanna, P., Arora, S., 2012. Silver nanoparticle-mediated enhancement in growth and antioxidant status of Brassica juncea. *Appl. Biochem. Biotechnol.* 167, 2225–2233.

Sharma, V.K., Yngard, R.A., Lin, Y., 2009. Silver nanoparticles: Green synthesis and their antimicrobial activities. *Adv. Colloid Interface Sci.* 145, 83–96. https://doi.org/10.1016/j.cis.2008.09.002

Siddiqui, M.H., Al-Whaibi, M.H., 2014. Role of nano-SiO2 in germination of tomato (Lycopersicum esculentum seeds Mill.). *Saudi J. Biol. Sci.* 21, 13–17.

Slomberg, D.L., Schoenfisch, M.H., 2012. Silica nanoparticle phytotoxicity to Arabidopsis thaliana. *Environ. Sci. Technol.* 46, 10247–10254.

Sondi, I., Salopek-Sondi, B., 2004. Silver nanoparticles as antimicrobial agent: A case study on E. coli as a model for Gram-negative bacteria. *J Colloid Interface Sci.* 275, 177–182. https://doi.org/10.1016/j.jcis.2004.02.012

Song, X., Sun, S., Zhang, W., Yin, Z., 2004. A method for the synthesis of spherical copper nanoparticles in the organic phase. *J. Colloid Interface Sci.* 273, 463–469. https://doi.org/10.1016/j.jcis.2004.01.019

Soni, M., Mehta, P., Soni, A., Goswami, G.K., 2018. Green nanoparticles: Synthesis and applications. *IOSR J. Biotechnol. Biochem.* 4, 78–83.

Spadafora, V., Orr, G., Wade, L., Wiglesworth, M., 2010. *Kasugamycin-A novel antibiotic for North American agriculture*. Presented at the Phytopathology, Amer Phytopathological Soc 3340 Pilot Knob Road, St Paul, MN 55121 USA, pp. S166–S166.

Speer, K.P., Quarles, L.D., Harrelson, J.M., Nunley, J.A., 1991. Tetracycline labeling of the femoral head following acute intracapsular fracture of the femoral neck. *Clin. Orthop. Relat. Res.* 267, 224–227.

Spotts, R., Cervantes, L., 1995. *Copper, oxytetracycline, and streptomycin resistance of Pseudomonas syringae pv syringae strains from pear orchards in Oregon and Washington*. Plant disease (USA).

Srilatha, B., 2011. Nanotechnology in agriculture. *J. Nanomed. Nanotechnol.* 2, 5.

Srinivasan, V., Nam, H.-M., Sawant, A.A., Headrick, S.I., Nguyen, L.T., Oliver, S.P., 2008. Distribution of tetracycline and streptomycin resistance genes and class 1 integrons in Enterobacteriaceae isolated from dairy and nondairy farm soils. *Microbial Ecol.* 55, 184–193.

Srivastava, S., Usmani, Z., Atanasov, A. G., Singh, V. K., Singh, N. P., Abdel-Azeem, A. M., Prasad, R., Gupta, G., Sharma, M., Bhargava, A. 2021. Biological nanofactories: Using living forms for metal nanoparticle synthesis. *Mini-Reviews in Medicinal Chemistry*. 21(2), 245–265.

Su, L.-X., Ma, X.-L., Zhao, K.-K., Shen, C.-L., Lou, Q., Yin, D.-M., Shan, C.-X., 2018. Carbon nanodots for enhancing the stress resistance of peanut plants. *Acs Omega* 3, 17770–17777.

Sullivan, T., Chapman, J., Regan, F., 2012. Characterisation of Nano-antimicrobial materials, in: *Nano-Antimicrobials*. Springer, pp. 181–208.

Sundin, G.W., Castiblanco, L.F., Yuan, X., Zeng, Q., Yang, C.-H., 2016. Bacterial disease management: Challenges, experience, innovation and future prospects. *Mol. Plant Pathol.* 17, 1506–1518. https://doi.org/10.1111/mpp.12436

Sundin, G.W., Wang, N., 2018. Antibiotic Resistance in Plant-Pathogenic Bacteria. *Annu Rev Phytopathol.* 56, 161–180. https://doi.org/10.1146/annurev-phyto-080417-045946

Suresh, S., Karthikeyan, S., Saravanan, P., Jayamoorthy, K., 2016. Comparison of antibacterial and antifungal activities of 5-amino-2-mercaptobenzimidazole and functionalized NiO nanoparticles. *Karbala International Journal of Modern Science* 2, 188–195.

Suslick, K.S., Hyeon, T., Fang, M., 1996. Nanostructured materials generated by high-intensity ultrasound: Sonochemical synthesis and catalytic studies. *Chem. Mater.* 8, 2172–2179. https://doi.org/10.1021/cm960056l

Tan, W., Hou, N., Pang, S., Zhu, X., Li, Z., Wen, L., Duan, L., 2012. Improved biological effects of uniconazole using porous hollow silica nanoparticles as carriers. *Pest Manag. Sci.* 68, 437–443.

Thenmozhi, M., Kannabiran, K., Kumar, R., Khanna, V.G., 2013. Antifungal activity of Streptomyces sp. VITSTK7 and its synthesized Ag2O/Ag nanoparticles against medically important Aspergillus pathogens. *Journal de mycologie médicale* 23, 97–103.

Tremblay, G.F., Bélanger, G., Lajeunesse, J., Chouinard, P.Y., Charbonneau, É., 2015. Timothy response to increasing rates of selenium fertilizer in Eastern Canada. *J. Agron.* 107, 211–220.

Tripathi, D.K., Shweta, Singh, Shweta, Singh, Swati, Pandey, R., Singh, V.P., Sharma, N.C., Prasad, S.M., Dubey, N.K., Chauhan, D.K., 2017. An overview on manufactured nanoparticles in plants: Uptake, translocation, accumulation and phytotoxicity. *Plant Physiol Biochem* 110, 2–12. https://doi.org/10.1016/j.plaphy.2016.07.030

Tso, T., Jeffrey, R., 1956. Effect of antibiotics on the growth and alkaloid production of tobacco plants. *Nature* 178, 800–801.

Vallad, G.E., Pernezny, K.L., Balogh, B., Wen, A., Figueiredo, J.F.L., Jones, J.B., Momol, T., Muchovej, R.M., Havranek, N., Abdallah, N., 2010. Comparison of kasugamycin to traditional bactericides for the management of bacterial spot on tomato. *Hort Sci.* 45, 1834–1840.

Van Overbeek, L.S., Wellington, E.M., Egan, S., Smalla, K., Heuer, H., Collard, J.-M., Guillaume, G., Karagouni, A.D., Nikolakopoulou, T.L., van Elsas, J.D., 2002. Prevalence of streptomycin-resistance genes in bacterial populations in European habitats. *FEMS Microbiol. Ecol.* 42, 277–288.

Vance, M.E., Kuiken, T., Vejerano, E.P., McGinnis, S.P., Hochella Jr, M.F., Rejeski, D., Hull, M.S., 2015. Nanotechnology in the real world: Redeveloping the nanomaterial consumer products inventory. *Beilstein J. Nanotechnol.* 6(1), pp.1769–1780.

Velmurugan, P., Lee, S.-M., Iydroose, M., Lee, K.-J., Oh, B.-T., 2013. Pine cone-mediated green synthesis of silver nanoparticles and their antibacterial activity against agricultural pathogens. *Appl. Microbiol. Biotechnol.* 97, 361–368.

Velmurugan, P., Shim, J., Kim, K., Oh, B.-T., 2016. Prunus× yedoensis tree gum mediated synthesis of platinum nanoparticles with antifungal activity against phytopathogens. *Mater. Lett.* 174, 61–65.

Verma, K.K., Song, X.-P., Joshi, A., Tian, D.-D., Rajput, V.D., Singh, M., Arora, J., Minkina, T., Li, Y.-R., 2022. Recent trends in nano-fertilizers for sustainable agriculture under climate change for global food security. *Nanomater.* 12, 173.

Vijayakumar, R., Koltypin, Yu., Felner, I., Gedanken, A., 2000. Sonochemical synthesis and characterization of pure nanometer-sized Fe3O4 particles. *Mater. Sci. Eng.: A* 286, 101–105. https://doi.org/10.1016/S0921-5093(00)00647-X

Wang, Y., Cui, H., Sun, C., Zhao, X., Cui, B., 2014. Construction and evaluation of controlled-release delivery system of Abamectin using porous silica nanoparticles as carriers. *Nanoscale Res. Lett.* 9, 1–6.

Wiener, P., Egan, S., Wellington, E., 1998. Evidence for transfer of antibiotic-resistance genes in soil populations of streptomycetes. *Molecular Ecology* 7, 1205–1216.

Wright, G.D., 2007. The antibiotic resistome: The nexus of chemical and genetic diversity. *Nat Rev Microbiol.* 5, 175–186. https://doi.org/10.1038/nrmicro1614

Xia, Z.-K., Ma, Q.-H., Li, S.-Y., Zhang, D.-Q., Cong, L., Tian, Y.-L., Yang, R.-Y., 2016. The antifungal effect of silver nanoparticles on Trichosporon asahii. J. *Microbiol. Immunol. Infect.* 49, 182–188. https://doi.org/10.1016/j.jmii.2014.04.013

Xue, J., Luo, Z., Li, P., Ding, Y., Cui, Y., Wu, Q., 2014. A residue-free green synergistic antifungal nanotechnology for pesticide thiram by ZnO nanoparticles. *Sci Rep.* 4, 5408. https://doi.org/10.1038/srep05408

Yang, W., Moore, I.F., Koteva, K.P., Bareich, D.C., Hughes, D.W., Wright, G.D., 2004. TetX is a flavin-dependent monooxygenase conferring resistance to tetracycline antibiotics. *J. Biol. Chem.* 279, 52346–52352.

Yuan, Z., Chen, Y., Li, T., Yu, C.-P., 2013. Reaction of silver nanoparticles in the disinfection process. *Chemosphere* 93, 619–625.

Yurkova, I.N., Omelchenko, A.V., Zubochenko, A.A., Panov, D.A., Danilova, I.L., 2020. Influence of a nanoseile on the contents of K+, Na+ and Cl– ions in wheat seeds in the conditions of the combined action of salting and drain. *Scientific Notes of the Crimean Federal University. Biol. Chem* 6(72), 220–230. https://doi.org/10.37279/2413-1725-2020-6-1-220-230

Zhang, L., Wu, L., Si, Y., Shu, K., 2018. Size-dependent cytotoxicity of silver nanoparticles to Azotobacter vinelandii: Growth inhibition, cell injury, oxidative stress and internalization. *PLOS ONE* 13, e0209020. https://doi.org/10.1371/journal.pone.0209020

Zhang, Z., Guo, H., Carlisle, T., Mukherjee, A., Kinchla, A., White, J.C., Xing, B., He, L., 2016. Evaluation of postharvest washing on removal of silver nanoparticles (AgNPs) from spinach leaves. *J. Agric. Food Chem.* 64, 6916–6922.

Zhao, G., Zhao, Y., Lou, W., Su, J., Wei, S., Yang, X., Wang, R., Guan, R., Pu, H., Shen, W., 2019. Nitrate reductase-dependent nitric oxide is crucial for multi-walled carbon nanotube-induced plant tolerance against salinity. *Nanoscale* 11, 10511–10523.

Zhu, H., Zhang, C., Yin, Y., 2004. Rapid synthesis of copper nanoparticles by sodium hypophosphite reduction in ethylene glycol under microwave irradiation. *J. Cryst. Growth.* 270, 722–728. https://doi.org/10.1016/j.jcrysgro.2004.07.008

9 Nanotechnology-based Antimicrobials in Agriculture

Thenmozhi Manivel, Sathammaipriya, Steffi P. F., and Thirumalaiyammal B.*
* Corresponding Author: steffi.mb@cauverycollege.ac.in

9.1 INTRODUCTION

Antimicrobials are used in agriculture to cure, control, and prevent disease, as well as boost output or growth. Antimicrobial use in plant agriculture is limited in terms of type and quantity due to financial constraints, a lack of antimicrobial efficacy for a variety of ailments, and environmental considerations. The US Environmental Protection Agency (EPA) oversees antibiotic usage in plants, while the Food and Drug Administration governs all other antimicrobial use. The Food Quality Protection Act of 1996 may eventually prohibit the use of antimicrobials in plant agriculture since the requisite reregistration and compliance with the higher criteria involved may be prohibitively expensive (Choudhary et al., 2023).

Since the use of antimicrobials on plants as well as their derivatives creates selective pressure on both targeted and non-target microbes, it is important to consider all of the effects of antimicrobial usage, antimicrobial resistance is an inevitable phenomenon, but it has been accelerated by the widespread and occasionally inequitable use of antimicrobials in industry, agriculture, and medicine. As a result, antimicrobial resistance has emerged as one of the most urgent public health issues in the world. As a result, various antimicrobial protection techniques are needed. Today, nanotechnology offers a reliable framework for modifying the physicochemical characteristics of a variety of materials to produce potent antimicrobials (Ansari, 2023).

With the use of nanotechnology, novel ideas and agricultural products have been developed that have a huge potential to solve the aforementioned issues. Although nanotechnology has made significant progress in pharmacology and medicine, agricultural applications have attracted much less attention. Current applications of nanotechnology in agriculture include controlled release of agrichemicals, delivery of plant hormones, seed germination, water management, transfer of target genes, and nanobarcoding. To use nanoparticles as protective agents or for precise and targeted administration via adsorption, encapsulation, and/or conjugation of an active, such as a pesticide, material scientists have engineered them to have desirable qualities, such as shape, pore size, and surface properties. The ability to produce a new generation of insecticides and other actives for managing plant diseases will significantly expand as agricultural nanotechnology improves (Songca, 2023; Prasad, 2014; Prasad et al., 2014, 2017).

In order to protect plants, nanoparticles can be used in a couple of different manners: either as the crop protection agent themselves, or as carriers for active substances that can be sprayed or soaked into the seeds, foliar tissue, or roots to protect them from pests. As carriers, nanoparticles can offer various advantages including (i) longer shelf life; (ii) higher solubility of weakly water-soluble pesticides; (iii) decreased toxicity; and (iv) increased site-specific uptake into the target pest. An additional advantage of nanocarriers might be an improvement in the activity and stability of

DOI: 10.1201/9781003268895-9

nanopesticides under environmental conditions (such as UV and rain), which would lead to a large decrease in the number of treatments and a consequent reduction in toxicity and the price of the pesticides. This chapter explores the halfway to facilitating green synthesis of nanoparticles and their usage in plant disease management, either on their own as protectants or as nanocarriers for pesticides like insecticides and fungicides (Daniel, Keyster, & Klein, 2023; Dhiman et al., 2022, 2023).

9.2 APPROACHES FOR SYNTHESIS OF NANOPARTICLES

Two fundamental processes, each involving different preparation techniques, are employed in the creation of nanoparticles, which are capable of being of artificial or natural origin and display distinctive features at the nanoscale. The "top-down" method, which is the first strategy, asks for breaking down dense substances into small bits by using external force. To supply the required energy for nanoparticle creation, this method makes use of a variety of physical, chemical, and thermal approaches. The second method, known as "bottom-up," relies on accumulating and fusing gas or liquid molecules or atoms.

Comparatively speaking, these two strategies offer benefits and drawbacks. Due to deficits and abrasion that can arise in nanoparticles, the up-down strategy, which is more expensive to apply, makes it hard to achieve ideal surfaces and edges. By contrast, the bottom-up approach can produce good results for nanoparticle synthesis. Furthermore, because the dimensions of the nanoparticles are better controlled with the bottom-up strategy, no waste materials need to be eliminated and smaller nanoparticles can be produced. Figure 9.1 lists the several categories of nanoparticle creation techniques. The top-down methodology's mechanical abrasion method grinds the material into particles using different ball mills and produces nanoscale alloys, composites, and semi-crystalline structures. The balls can contaminate the process even though it is inexpensive, effective, and effortless (Nadaroglu et al., 2017).

9.3 GREEN SYNTHESIS OF NANOPARTICLES

Conventional chemical and physical approaches require the reduction of metal ions and the employment of stabilizing agents to prevent the unwanted agglomeration of the created nanoparticles. These processes present a risk of toxicity to the cell and the environment. Additionally, the surface chemistry, size, shape, and composition of the generated nanoparticles are considered to be hazardous. Environmentally safer synthesis techniques are required as a result of the pollution that traditional synthesis techniques cause.

The world of research is relentlessly looking for answers to stop the destruction caused by harmful manufacturing practices as the effects of atmospheric conditions become more widespread. Natural biological systems are used to produce nanomaterials using green synthesis techniques. By combining natural starting materials and low-energy procedures, green synthesis is a technique that is equally as effective as traditional synthesis, if not more so. It offers a sustainable means of producing nanomaterials. Active compounds have recently been used in endogenous life forms like bacteria, yeast, algae, and fungi to successfully synthesize diverse nanoparticle systems. Green synthesis is thus a practical solution to the problems with traditional synthesis methods in research and industrial production (Hussain et al., 2023).

Bacteria are obvious targets in the manufacture of nanoparticles due to their quick development, low cost of culture, and ease of regulating and manipulating the growing environment. In addition, it is well known that several bacteria species have unique defences against the adverse effects of metals or metallic substances. Metal ions are capable of being diminished and precipitated for the creation of nanoparticles using mechanisms of metabolism and reducing agents found in bacteria, including proteins, enzymes, etc. (Prasad et al., 2016) While some

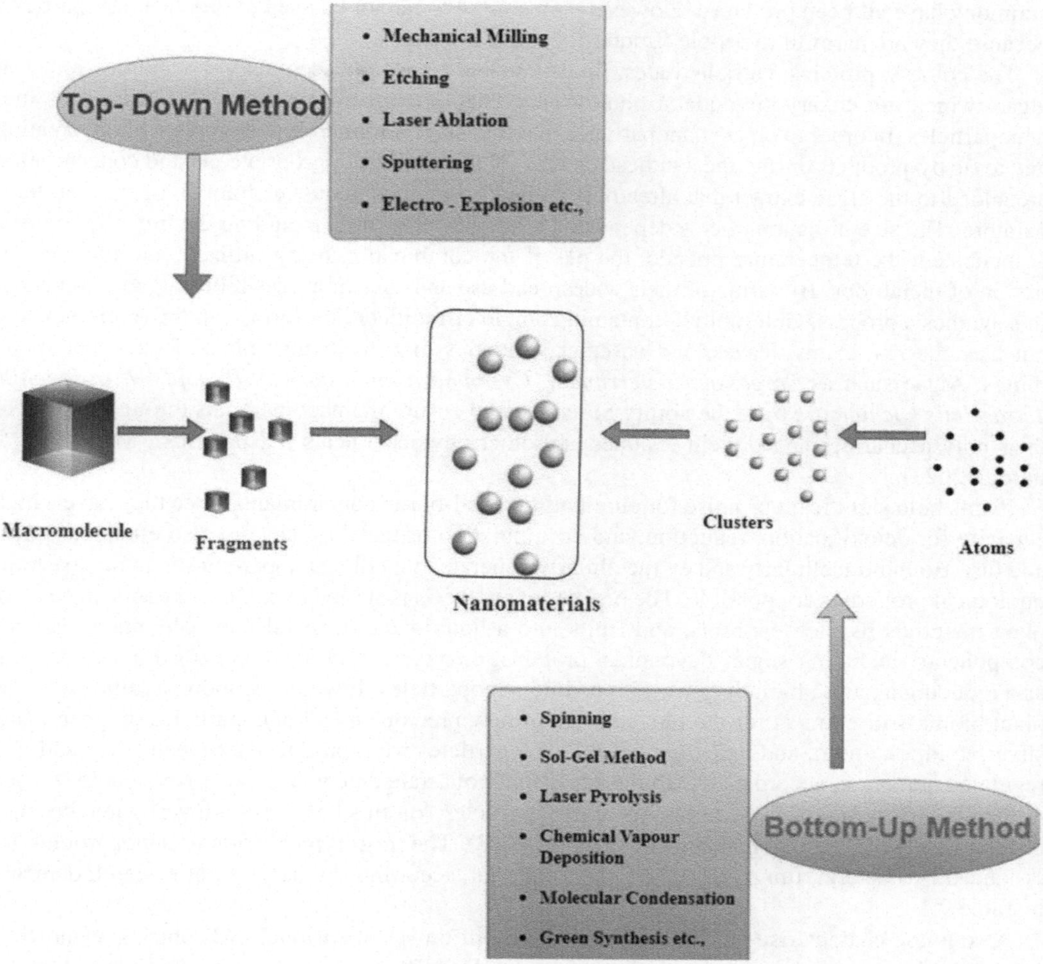

FIGURE 9.1 Categories of nanoparticle generation.

bacterial strains can be extremely harmful, many others are found inevitably in the body and present little to no risk to anyone dealing with them. Additionally, a lot of these strains, including *Bacillus subtilis* and *Escherichia coli (E. coli)* are simple to culture and have easily modifiable genetic codes. These properties make the process of nanoparticle manufacturing in bacteria possible to implement (Kumari et al., 2023).

Utilizing easily-cultured, quickly reproducing eukaryotic yeasts and moulds with simple biomass design, one can produce nanoparticles either extracellularly or intracellularly using enzymes. The metallic ion concentrations employed and the incubation conditions have an impact on the dimensions of the nanoparticles created. *Saccharomyces cerevisiae*, a species that breaks down carbohydrates into alcohols and carbon dioxide, is the most traditional and widely used type of yeast. Utilizing the fermentation process, this particular species is employed in baking and the production of alcoholic beverages. The filamentous fungi such as *Trichothecium* and *Verticillium* spp., also have different nanosystems than those produced by bacteria and yeast. Utilizing different fungal species, nanoparticles of silver, gold, cadmium sulfide, lead sulfide, ferrous oxide, selenium, and antimony have all been produced. However, some moulds cannot be used to produce nanoparticles because they are harmful to people (Luque-Jacobo et al., 2023).

The colours, proteins, carbohydrates, lipids, nucleic acids, and secondary metabolites found in algae, which are eukaryotic aquatic phototrophs, enable them to break down metallic ions into nanoparticles. In order to create nanoparticles that may have antimicrobial features without creating any toxic by-products during the synthesis, metal solutions of the appropriate pH and concentration are added to the algae extract that already is embedded in an aqueous medium at a particular temperature. The size of nanoparticles depends on a number of factors, including the time the solution is incubated, the temperature outside, the pH of the combination being utilized, and the concentration of metal ions. By virtue of their widespread use and ease of accessibility, algae also benefit this synthesis process. Chlorophyll-containing single or multicellular forms can thrive on moisture but lack the real stems, leaves, and vascular structures that distinguish plants from other living things. Algae such as *Sargassum tenerrimum, Cystophora moniliformis, Pithophora oedogonia, Tetraselmis kochinensis* have the ability to reduce and ensure the nanomaterials through the use of their polysaccharides and protein residues and other phytochemicals (Alshammari, Mahmoud, & Farrag, 2023).

Plants hold significant promise for eliminating metal-borne contaminants since they have a high capacity for detoxification, reduction, and accumulation of metals. They are also efficient, quick, and fast. Both intracellularly and extracellularly generated metallic nanoparticles with diverse morphological properties are possible. The process of synthesis is started by adding extracts made from plant materials like leaves, roots, and fruits into a liquid state of metal ions. The plant extract's components, including sugar, flavonoids, proteins, enzymes, polymers, and organic acids, work as a reducing agent to bioinduce metal ions into nanoparticles. It would be more significant to use plant biomass or extracts for the biosynthesis of new precious metal nanoparticles (copper, Zinc, silver, gold, platinum, and palladium) if the nanoparticles were produced extracellularly and in a regulated manner in accordance with their dispersity of arrangement and size. Plants could be used in the manufacturing of these noble metal nanoparticles, but this hasn't been investigated because of their diverse biological makeup (Peng et al., 2023). The most typical sorts of nanoparticles are produced by bacteria, fungi, algae and plant material, according to the relevant research depicted in Table 9.1.

A common characteristic of NPs is the existence of three-dimensional (3D) objects. Nanowires and nanotubes are examples of one-dimensional (1D) NPs, while nanolayers and nanofilms are examples of two-dimensional (2D) NPs, which have one dimension at the nanoscale and two dimensions at the macroscale. Again, 3D NPs (nanoballs, nanoflowers) have 3 dimensions at the macroscale but 0 dimensions at the nanoscale, whilst zero-dimensional (0D) NPs have all 3 dimensions at the nanoscale. In order to create zero-dimensional NPs with precisely controlled dimensions, for example, a wide range of physical and chemical approaches have been devised (Tiwari et al., 2012). In light-emitting diodes, solar cells, and single-electron transistors used in lasers, zero-dimensional nanoparticles (NPs) such as quantum dots are widely accepted and used. Research in nano-engineering has increasingly focused on the production of two-dimensional NPs, including junctions (continuous islands), branched structures, nanoprisms, nanoplates, nanosheets, nanowalls, and nanodisks. An explosion in the exploration and creation of intriguing applications in sensors, photocatalysts, nanocontainers, and nanoreactors has been caused by the geometric structures of NPs. The enormous extent of the surface and other excellent qualities of three-dimensional NPs, however, has recently gained significant study interest. These attributes include absorption destinations for all implicated molecules in a limited region. As a result, the development and refinement of novel technologies for the manufacture of NPs and their prospective applications are of great importance, especially for the creation of environmentally and agriculturally sustainable systems (Shang et al., 2019).

TABLE 9.1
Nanomaterials and Their Biological Sources

S.No	Source	Organism
1.	Bacteria	*Aeromonas*
		B.cereus
		B.subtilis
		Desulfobacteriaceae
		Clostridium spp.,
		E.coli
		E.coli DH5α
		Lactobacillus spp.,
2.	Fungi (Yeast & Mould)	*Aspergillus fumigatus*
		A.oryzae
		A.niger
		Rhizopus spp.
		Candida spp.
		Candida utilis
		Yarrawia spp.
		Penicillium spp.
3.	Algae	*Sargassum plagiophyllum*
		Ulva fasciala
		Portieria hornemanni
		Jania rubens
		Desmodesmus abundans
4.	Plants	*Cassia auriculata*
		Trachyspermum amni
		Carissa carandas
		Sisymbrium irio
		Clerodendrum inerme
		Cinnamomum verum
		Cymbopogom citratus

9.4 PHYSICAL AND CHEMICAL PROPERTIES OF SYNTHESIZED NANOPARTICLES

Three layers make up a nanoparticle: the shell, the core, and the top layer. A wide range of molecules, including metal ions, surfactants, and polymers, often make up the surface layer. One material may make up a nanoparticle, or it may be made up of numerous different components. In accordance with their chemical and electromagnetic characteristics, nanoparticles are feasible as suspensions, colloids, or diffused aerosols.

How a nanoparticle behaves depends on its size. The malleability and ductility of bulk copper are not present in copper nanoparticles that are smaller than 50 nm since they are extremely hard materials. Additionally, magnetic materials' superparamagnetic, semiconductor Q-particles' quantum confinement and certain metals' plasmon resonance on the surface are all influenced by the size of nanoparticles.

Additionally, studies have shown that nanoparticles absorb solar energy in photovoltaic cells far more efficiently than thin films made of bulk material's continuous sheets do. This is so that they can more effectively absorb solar radiation due to their reduced size.

The large surface area to volume ratio of nanoparticles causes them to diffuse more readily at high temperatures. Sintering can occur at lower temperatures with nanoparticles than with bigger particles thanks to this characteristic. Nanoparticles' ability to diffuse may not change the product's density, but it can cause agglomeration.

At the nanoscale, nanomaterials' chemical characteristics also alter. Because nanoparticles have a higher percentage of surface atoms than bulk objects, their reactivities are higher than those of bulk materials. Among the chemical characteristics are some of the following: (i) one of the main factors influencing how materials behave at the nanoscale is the predominance of surface. Nanoparticle features like electrical transport are no longer governed by solid-state bulk phenomena since as much as 50% of the atoms that exist in nanoparticles are surface atoms. (ii) Due to the increased percentage of surface atoms in nanomaterials, their average energy is higher than that of atoms in longer structures. For instance, when the catalyst gets smaller at the nanoscale, the amount of chemical activity per exposed atom also increases. (iii) Surfaces and interfaces may operate as magnets for imperfections and impurities, and associations among particles in those small dimensions may be influenced by the type of chemical bonding and surface structure. (iv) surface properties and the mediating of interactions between nanoparticles can both be altered or controlled using molecular monolayers.

9.5 NANOPARTICLES' ANTIMICROBIAL ACTIVITY ON PHYTOPATHOGENS

Plant diseases are controlled by a wide range of synthetic and organic products. To manage phytopathogens, which include fungi, oomycetes, viruses, bacteria, and nematodes and are all to blame for over 70% of crop diseases, chemical pesticides are frequently utilized. However, the majority of these synthetic items are derived from petrochemicals and pose a health risk since they may assemble in the soil and aquifers. In addition, using pesticides, fungicides, and herbicides results in additional issues such as reduced insect pollination, injury to domestic animals, the development of pathogenic microbial resistance, and ravage tolerance to various chemical composites (Álvarez et al., 2016).

Pests and plant diseases have a significant negative impact on agricultural productivity globally, leading to billion-dollar economic losses and lowering food quality. The World Health Organization (WHO) also conducted studies in 2010 that revealed farmers' lack of awareness of the toxicity of broad-spectrum pesticides, their heavy reliance on them, and poor compliance with regulatory norms. The regulation of pesticides and their safe management is severely hampered by these dangerous practices. Millions of tonnes of pesticides are currently used in agriculture, with industrialized countries using 20% of the world's supply and being responsible for 99% of all pesticide-related fatalities (Hernández-Díaz et al., 2021).

Resistance of phytopathogens to the chemical products used to control them is another issue that is progressively gaining prominence. This resistance is primarily caused by the excessive and ongoing use of agrochemicals without adhering to the manufacturer's instructions or any applicable laws on the dosage and toxicity of these goods. Economic losses, environmental harm, and deteriorating human health are the results of this non-compliance. In accordance with the Food and Agriculture Organization of the United Nations (FAO), during the past 50 years, pesticide sales have increased significantly worldwide, with annual spending now topping 35 billion USD. As a result, creating substitute products is a useful crop protection tactic. NPs are a new option that may have antibacterial activity that has recently come to light. This result is ascribed to the nanoparticles' tiny size and high surface-to-volume ratio, which facilitate their contact with phytopathogen membranes (Mateo-Sagasta et al., 2018).

Several soil-borne infections have been successfully combatted by nanoparticles, which can be used alone as antimicrobial agents (Mubeen et al., 2023). It can be used to guard against pests and pathogens like fungi, bacteria, and viruses by applying it to soil, seeds, roots, and leaves. Nanoparticles enter the plant's system and either directly combat pathogens or act as elicitor molecules to trigger both local and systemic defines responses in plants. The most extensively investigated nanoparticles are those made of metals, such as gold, silver, titanium, zinc, and copper oxide, which are also known to have antifungal, antibacterial, and antiviral characteristics (Gupta et al., 2018; Ismail et al., 2017; Zhang et al., 2023).

9.6 MODE OF ACTION AGAINST PHYTOPATHOGENS

Several mechanisms have been identified to date, including protein dysfunction such as oxidation of cysteine in the Fe-binding site, destruction of the Fe-S cluster (like exchange of catalytic metal, as well as the exchange of structural metal), the production of reactive oxygen species (ROS) and antioxidant depletion, compromised functioning of membranes (including membrane damage, loss of membrane potential), disturbance with nutrient uptake (due to inhibition of the Fe (III) transporter gene expression), and these defences against different phytopathogens may not work separately, but they do so in combination (Caparco, Gonzalez-Gamboa, Hays, Pokorski, & Steinmetz, 2023).

The electrostatic interaction that exists between the negatively charged membranes of cells of bacteria and NPs with either positive or low negative charges causes NP adherence to the microbial cell membrane. The NPs affect the morphological configurations of the membrane, and membrane depolarization disrupts the permeability of the membrane and respiratory activities, ultimately causing cell death (Aziz et al., 2014, 2015, 2016, 2019). Due to this breakage, the evacuation of secretions such as enzymes, protein molecules, DNA, and metabolites occurs. Furthermore, the NPs may generate irregular pits in the microbial cell wall, which facilitates NP passage into the periplasmic region and inside the cells (Sela et al., 2023) (Figure 9.2 and Table 9.2).

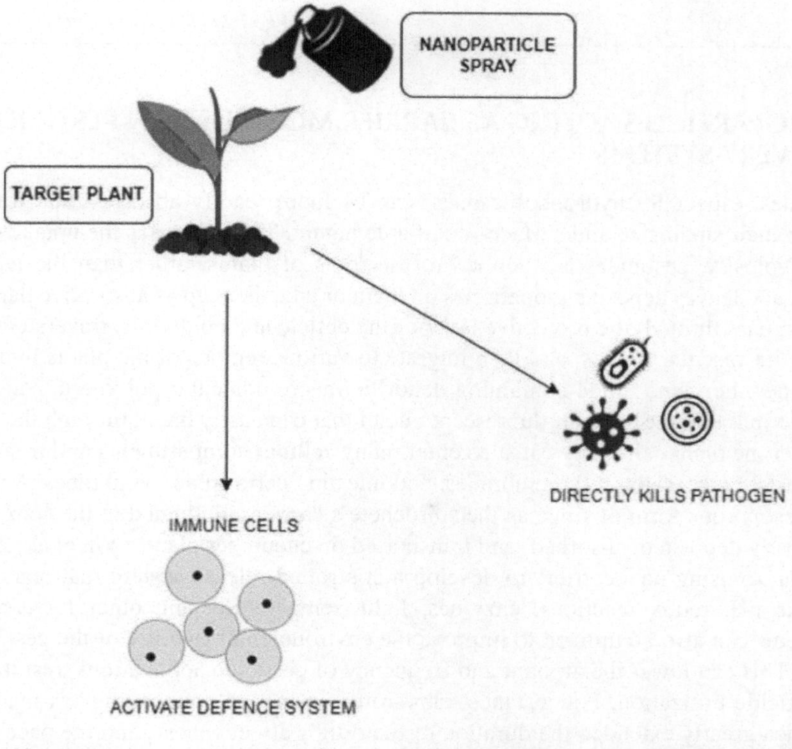

FIGURE 9.2 Nanoprotectants spray induces immune system in indirect mechanism as well as directly acts on pathogen.

TABLE 9.2
Nanoparticles Acting against Phytopathogens

S.No	Nanoparticles	Target Pathogens
1.	Silver	*Collectrotrichum gloeosporioides* *Alternaria solani* *Fusarium oxysporum* *Fusarium solani* *Macrophomina phaseolina* *Rhizoctonia solani* *Aspergillus niger* etc.,
2.	Copper-based nanoparticles	*Xanthomonas oryzae* pv. *Oryzae* *Xanthomonas campestris* pv. *phaseoli* *Alternaria solani* *Fusarium solani* *Aspergillus flavus* etc.,
3.	Carbon-based nanoparticles	*Aspergillus niger* *Aspergillus oryzae* *Fusarium oxysporum*
4.	Chitosan nanoparticles	Alfalfa mosaic virus, *Fusarium* spp., *Botrytis* spp., *Pyriularia grisea*
5.	Zinc oxide	*Aspergillus flavus*
6.	Magnesium oxide	*Xanthomonas oryzae* pv. *Oryzae*
7.	Nickel	*Collectrotrichum musae*
8.	Gold	*Puccinia graminis pv. tritci*

9.7 NANOPARTICLES ACTING AS CARRIER MOLECULES IN PESTICIDE DELIVERY SYSTEMS

Nanopesticides, especially hydrophobic ones, can be more readily absorbed and transported by plants due to their small size and surface-modifiable nature. This increases the uptake of pesticides by plants. Typically, pesticides are applied to the roots of plants rather than the leaves. Plants' stems, roots, and leaves deposit nanoparticles on them or take them up as adsorptive particles. These particles then pass through the protective layer of the cuticle and epidermis, traverse the plastids or exoplasts to the vascular tissues, and then migrate to various regions of the plants by means of the vascular tissues. Perylene imide-containing dendritic macromolecular polymeric nanocarriers can self-assemble into complexes with fluorescent nuclei that can easily break through the root cap cell wall and enter the plant cell. They can also enter many cellular compartments and tissues of insects as well as pests' body walls, cell membranes, and intestine peritrophic membranes. As an example, using the mesoporous form of silica as the spirochete's carrier culminated in the spirochetes being more effectively deposited, absorbed, and transported on cucumber plants (Yin et al., 2023).

In addition to using nanocarriers to develop a controlled-release system that reacts to external elements like pH, redox reactions, enzymes, light, temperature, and other factors, controlled-release systems can also be utilized to improve the environmental stability of the pesticides' active ingredients. This can lower the amount and frequency of pesticide applications, resulting in higher rates of pesticide utilization. For instance, chrysromycin-loaded nanoconjugates made of amino-modified silica greatly extended the duration of the drug's effectiveness, and the pace at which the active ingredient was released was influenced by temperature, the pH of the environment, and particle size (Ahmed, Saleem, & Hassanzadeh, 2023).

9.8 NANOPARTICLES ACTING AS CARRIER MOLECULES IN HERBICIDE DELIVERY SYSTEMS

Herbicides are applied to crops in a variety of ways, such as (i) pre-sowing treatment, which involves treating the soil with herbicide before the sowing or cultivation of the crop; (ii) pre-emergence administration, which involves applying herbicide after sowing or planting the crop but before the crop emerges; and (iii) post-emergence treatment, which involves applying herbicide after the crop plants have fully emerged; (iv) in wide-spaced-row crops, direct application involves spraying broad-spectrum herbicides throughout the space between the rows with no alleviating the crop fallen leaves or crop shoots; (v) band application involves spraying herbicides only over the crop rows while leaving the interrow spaces untreated. The most common approach of application for cost-intensive herbicides is typically soil treatment. Although herbicides are used to eradicate weeds, they can also have a significant negative impact on food crop production and potentially tamper with consumers. The emergence of nanoparticle-based herbicides is a consequence of scientists searching for better alternatives to currently used herbicides in order to successfully eradicate weeds with little to no harm to cultivated crops (Linh, Qui, & Triatmojo, 2022).

To ensure better weed eradication and increased bioavailability, herbicides are coated on nanomaterials (NMs). It is necessary to find the ideal relationship with their nanostructure and herbicide components in order to create the NM, which could load and distribute considerable quantities of herbicide to seeds of herbs within the soil throughout the non-crop stage or into living plants. The active substance is attached to, adsorbed onto, enclosed within, or trapped inside the nano-matrix. Due to the slow-release properties of NMs, the contents' attachment to the material, and environmental factors, sustained discharge of the active component is accomplished. Their area of coverage, higher solubility, higher mobility, and decreased toxicity due to the removal of organic solvents are all advantages of NM-based formulations (Sindhu et al., 2023).

Herbicide delivery systems that effectively reduce weeds while protecting non-target organisms have been the subject of several investigations. Unfortunately, some of the most exciting nanostructured approaches that can be combined with herbicides have been centred on polymeric materials. Chitosan, a biodegradable material produced by the deacetylation of chitin, has been singled out for application in the agricultural sector. Because chitosan nanoparticles are poorly soluble in aqueous solutions, they are mixed with other organic, inorganic, or polymeric substances to make them more soluble. These substances adhere to stems as well as leaves epidermis, which allows them to exert their effects at the target areas for a longer period of time. The nanoparticles' characteristics can be improved by having an abundance of amines and hydroxyl groups. It is a highly effective carrier system for agricultural commodities and plant micronutrients (Kaushal et al., 2023).

The unique shapes, dimensions, and structure of silica-based nanoparticles enable them to carry active substances to the target site. Examples of them include mesoporous silica nanoparticles (MSNs) and heterogeneous hollow silica nanoparticles (PHSNs), which are primarily spherical or circular in shape with holes that resemble pores. To keep them safe and ensure effective release at target places, the active chemicals are mostly placed in the inner hole. The outer shell protects the active components from UV radiation deterioration (Teodoro et al., 2023) (Figure 9.3).

9.9 NANOMATERIALS IN EMULSION FORMS

Nanoemulsions, which are further referred to as submicron emulsions, ultrafine emulsions, or mini-emulsions, are economically and structurally stable kinetic isotropic dispersions of submicron-sized colloidal particle-like systems composed of two liquids that are immiscible (e.g., water and oil) centred by surfactant and co-surfactant forming a single phase in the form of small droplets

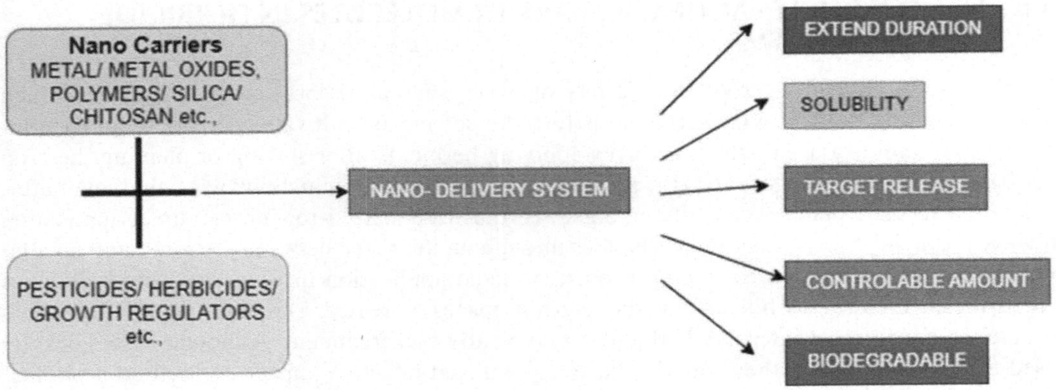

FIGURE 9.3 Advantages of using nanocarriers.

0.1–100 m in diameter. At times nanoemulsions have been divided into three groups: bicontinuous (in which microdomains of water and are interdispersed within the system), oil-in-water type, and water-in-oil type (in which oil is distributed in the aqueous phase). Oil-in-water and water-in-oil emulsions are both present concurrently in one system in multiple emulsions, another type of nanoemulsion (Malode et al., 2021).

In comparison to other dosage forms, nanoemulsions have a number of benefits, including (1) faster rate of absorption; (2) lower variability in absorption; (3) protection from oxidation and hydrolysis in O/W nanoemulsions; (4) delivery of lipophilic drugs after solubilization; (5) aqueous dosage form for water-insoluble drugs; (6) increased bioavailability for many drugs; (7) ability to incorporate both lipophilic and hydrophilic drugs; and (8) control of drug release via permeation through a liquid film, whose hydrophilicity or lipophilicity and thickness can be carefully adjusted (Gurpreet et al., 2018).

9.10 METHODS OF GENERATING NANOEMULSIONS

The intermediate layer formed by surfactant and co-surfactant stabilizes both the continuous and dispersed phases of a colloidal nanoemulsion, which is a continuous nanosystem. Oil and water, which are nanoemulsions' primary components and are incompatible with one another, require both mechanical and chemical energy to create. Low-energy emulsification methods use chemical energy emanating from the emulsion system, while high-energy emulsification methods use mechanical energy to create nanoemulsions. High-pressure homogenization, high-shear mixing, and ultrasonication are all components of the high-energy emulsification technique. Phase inversion temperature, phase inversion composition, also known as the emulsion inversion point, and solvent displacement are all components of the low-energy emulsification method (Sharma et al., 2020) (Figures 9.4 and 9.5).

9.10.1 High-pressure Homogenization

This method creates nanoemulsions from tiny droplets using a high-pressure homogenizer, also known as a piston homogenizer. Intense hydraulic shear is produced as a result of the mixture being forced at a high pressure of 500 to 5000 psi via a small inlet orifice, resulting in extremely smaller particle sizes. For the combination of laboratory and large-scale applications, this preparation technique is effective.

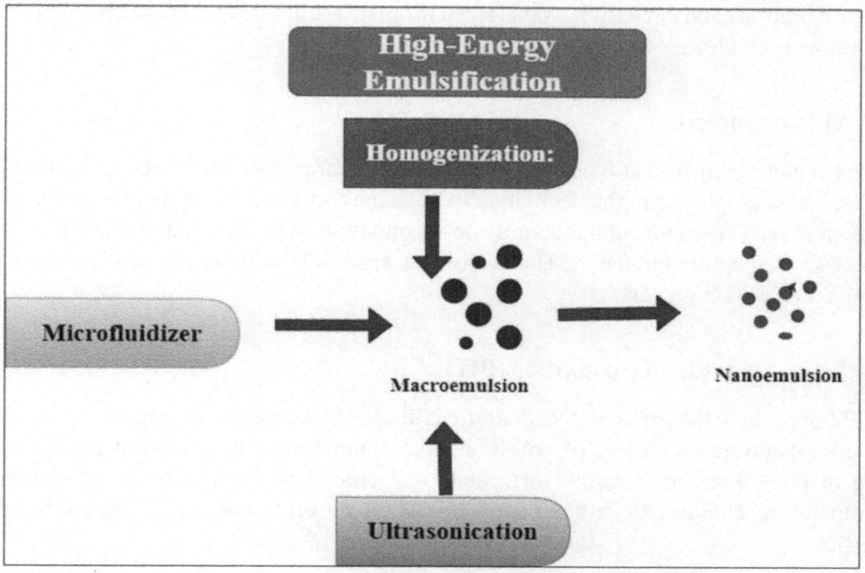

FIGURE 9.4 Exemplification of nanoemulsion preparation – high-energy methods.

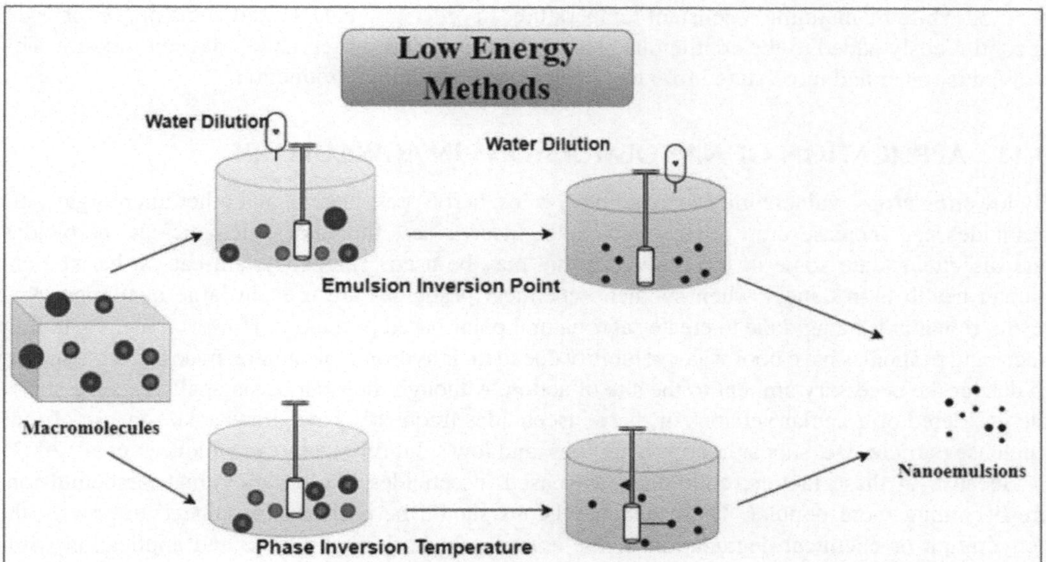

FIGURE 9.5 Exemplification of nanoemulsion preparation – low-energy methods.

9.10.2 Ultrasonication

Due to its simplicity and cost-effectiveness, ultrasonication is the most popular method of creating nanoemulsions, particularly for laboratory research. When the sonicator's tip makes contact with the formulation, the ultrasonic generator, which generates frequencies more than 20 kHz, causes mechanical vibrations and cavitation. Nano-sized droplets are created as a result of such disruptions, which combine both phases.

9.10.3 Microfluidizer

Using an extremely high-pressure positive displacement pump (500–20,000 psi), a device known as a microfluidizer is employed in this technique to push the formulation via tiny channels and into the impingement area. The enduring nanoemulsion formulation with the desired-sized nanodroplets is then created from the bulk emulsion. The use of this approach is limited to small-scale and laboratory settings due to its high cost.

9.10.4 Phase Inversion Temperature (PIT)

With the PIT approach, the surfactant's curvature is altered by keeping its composition constant while raising the temperature. A variety of non-ionic surfactants with temperature-dependent solubility, consisting of polyethoxylated amine surfactants, are crucial in the generation of nanoemulsions. Rapid temperature changes (a reduction of 25–30°C) prevent coalescence and result in steady nanoemulsion.

9.10.5 Emulsion Inversion Point (EIP)

Another name for Phase Inversion Composition (PIC), this technique involves changing the composition while maintaining a constant temperature. In order to create a nanoemulsion, oil or water is continuously added to the continuous phase, which is an oil, water, and surfactant mixture. This method is easier and more suited for large-scale nanoformulation production.

9.11 APPLICATION OF NANOEMULSIONS IN AGRICULTURE

By lowering crops' vulnerability to invading species, herbivores, insects, and other microorganisms, pesticides can increase crop quality and yield. Insecticides, fungicides, nematicides, herbicides, and insecticides are some of the pesticides that may be used. The ecosystem can be harmed and human health is in danger when synthetic chemical pesticides are used in large quantities. As a result, research is being done to create safer natural plant-based pesticides. However, many naturally occurring pesticides have poor water solubility due to their hydrophobic nature, making it challenging to deliver the necessary amount to the site of action. Although they can occasionally become solids, the powdered or granular versions of these insecticides frequently have drawbacks because of their immense particle size, subpar adhesion abilities, and low solubility features (Mulqueen et al., 2003).

Because of these factors, colloidal plant-based insecticides like antimicrobial nanoemulsions are becoming more popular. These nanoemulsions should be properly formulated to prevent the dehydration or chemical degradation of the antimicrobials during storage and application while still being effective against the intended spoilage or infectious microorganisms. The strength and/or dosage of plant-based antibiotics can be increased or decreased with the use of nanoemulsions, potentially lowering their possible toxic effects. A hexane-soluble sample of *Manilkara subsericea*, for instance, was used to create insecticidal nanoemulsions that were used to combat the *Dysdercus peruvianus*, a significant cotton pest. Recent research has shown that nanoemulsions were efficient at eradicating pests while also posing less of a threat to the environment and unintended creatures. Neem oil-based nanoemulsions have been demonstrated to be enforceable against two insects that damage significant commercial crops, the red flour beetle and the rice weevil (*Sitophilus oryzae* and *Sitophilus oryzae*, respectively). The nanoemulsion droplet size had an impact on the incidence of mortality, with minuscule particles being more efficient. It has been demonstrated that pesticides made from *Mentha piperita* oil work well against cotton aphids. An eco-friendly and secure plant-based pesticide could be made using nanoencapsulated *Melissa officinalis* L. oil, which has been shown to have high activity as an insecticide against *Tribolium castaneum* Herbst. According to these findings, greenly synthesized nanopesticides may be a good substitute for pesticides that are made of chemicals (McClements et al., 2021).

9.12 NANOMATERIALS IN PLANT-GROWTH-PROMOTING ASPECTS

The community to which a plant belongs has a variety of factors that affect its ability to survive. Plant-growth-promoting bacteria (PGPB) are known to improve plant health through a variety of direct and indirect ways. The direct mechanisms involve the creation or control of hormones associated with plant growth, such as auxin, cytokinin, or gibberellin, as well as the uptake of vital nutrients from the soil, such as nitrogen, phosphorus, and iron. Indirectly, PGPB reduces the availability of nutrients for pathogens, synthesizes pathogen-inhibiting volatile compounds, induces systemic resistance in the plant, and helps plants withstand biotic stress by producing antibiotics, antioxidant enzymes, and other molecules. Only 30–50% of chemical fertilizers can be absorbed by crop plants, therefore a sizeable portion remains in the soil and contaminates groundwater. Due to saturation, fertilizer effectiveness has thus diminished over time. Live or dormant microorganisms (inoculants) are the main component of biofertilizers, which operate as a delivery system for microbes to boost nutrient availability for plants. The formulation of biofertilizers allows for easy handling and extended storage. The use of PGPB as inoculants in agriculture is widespread throughout the world. Examples include nitrogen fixers, phosphorus solubilizers, potassium solubilizers, and biocontrol agents (Srivastava et al., 2016).

The market for biofertilizers is dominated by nitrogen-fixing inoculants since plants cannot transform air nitrogen into usable nitrogen, a nutrient that is necessary for plant vitality. The most often employed inoculant species belong to the genera *Acetobacter, Azotobacter, Azospirillum*, and *Pseudomonas*. PGPB are promising and are already being utilized commercially as inoculants in cropping systems, but due to issues with their stability in the soil, field applications, and delivery systems, their practical application may not correspond with the projected performance in plants. In general, bacterial populations fall quickly after inoculation, reducing bacterial activity in the rhizosphere. This is because numerous factors influence bacterial colonization in plants. Both a suitable microenvironment and sustained physical protection are required for PGPB in order to stop this deterioration in the field. These factors make peat or liquid carriers necessary when using PGPB in agricultural systems since they help to stabilize and sustain the bacteria while they are being transported and stored (Yang et al., 2013).

To protect the quality and efficacy of the bacterial inoculum, as well as to cut production costs and negative effects, alternative materials must be investigated. In order to achieve these goals, nanostructured materials might be a cutting-edge and effective strategy that improves the stability and targeting of beneficial microbes in agriculture (Osherov et al., 2023). By enhancing their resilience to desiccation, heat, and UV radiation, the addition of nanoformulations can produce biofertilizers that are stable and repeatable. By incorporating microbial cells with inorganic or organic NPs, this method delivers the bacteria to the target tissues at predetermined periods while maintaining controlled conditions. Irrespective of the NP substance, the huge specific surface area of NPs with a negative charge and various groups largely dictates the specificities and interactions between microorganisms and NPs. Despite having a net negative charge, solid particles and cells may have positive charges and hydrophobic areas that let microbes and nanomaterials adhere to one another. Because NPs can adhere to hydrophobic patches already present on bacteria or because they can produce hydrophobic regions on bacterial cell membranes, aggregates and binding patterns may develop. But interactions between particles and bacteria don't just happen because of electrostatic attraction; they can also happen because of chemical reactions on the bacterial surface, such as those brought on by exposure to the phospholipid membrane. Perhaps the most significant biomolecules interacting with NPs are the phospholipids, proteins, lipopolysaccharides (LPS), and lipoteichoic acid (LTA) of bacterial cells. There is still no specific model to anticipate PGPB and NP interactions. Instead, the manner of NP transport into bacterial cells depends on the size of the molecules (Timmusk et al., 2017).

Nanomaterials have a positive effect on beneficial bacterial features like nitrogen fixation and the synthesis of secondary metabolites. For instance, zinc oxide NPs boosted siderophore formation, most likely driven by ion release in the plant growth-promoting bacteria *Pseudomonas chlororaphis*, while copper oxide NPs increased IAA synthesis in the bacterium. By boosting the expression of the nitrification-related genes, silver nanoparticles (AgNPs) raised the nitrogen fixation capability in *Nitrosomonas europaea*. These investigations also revealed the detrimental effects of NPs on advantageous bacterial characteristics, demonstrating the unpredictability and dependence of NPs' effects on bacteria. It is true that there are a number of processes by which nanomaterials affect bacterial metabolism; however, these mechanisms are not fully understood. Ion interactions, changes in gene expression, and modifications to cell membranes are reportedly among these methods (de Moraes et al., 2021).

9.13 CONCLUSIONS AND FUTURE PROSPECTS

In the realm of future agriculture, the integration of nanotechnology-based antimicrobials marks a groundbreaking concept with the potential to revolutionize food production. By leveraging the precision and versatility of nanomaterials, this innovation offers a sustainable solution to the challenges posed by microbial infections in crops. Nanotechnology enables the design and fabrication of nanosized structures with tailored properties, facilitating the development of antimicrobial agents that can target specific pathogens while minimizing ecological impact. These nanomaterials could be engineered to release controlled doses of antimicrobial compounds directly to plant surfaces, enhancing their effectiveness and reducing the need for conventional, often environmentally harmful, chemical treatments. This concept not only holds the promise of increased agricultural yields but also aligns with the growing demand for eco-friendly and efficient farming practices, paving the way for a more resilient and secure global food supply chain.

Not many nanoparticle-based solutions have been commercialized for agricultural application, despite the numerous potential benefits connected with the usage of nanoparticles. Numerous elements, including a dearth of field experiments and underuse of pest-crop host systems, could account for the lack of commercial applications. Nanotechnology has advanced quickly in other industries, and in order to keep up with it for agricultural applications, it is essential to comprehend the fundamental research questions and fill in any knowledge gaps. The use of NMs in agrochemicals is still being investigated, and concerns about public safety, production costs, evaluation standards, registration procedures, and cost must all be taken into consideration. Before nanoparticles are deemed safe for usage, their effects must be thoroughly evaluated.

ACKNOWLEDGMENTS

Thenmozhi Manivel[1], Sathammaipriya[2], Steffi P F*[2], and Thirumalaiyammal B[3] were involved in the conception of the review and the authors also wish to acknowledge the reviewers who provided helpful comments on the report, as well as for help and advice throughout this review article and thank our institution for more support.

REFERENCES

Ahmed, S. A., Saleem, M. F., & Hassanzadeh, H. (2023). Optimization of solvent evaporation method in liposomal nanocarriers loaded-garlic essential oil (Allium sativum): Based on the encapsulation efficiency, antioxidant capacity, and instability. *IET.Nanobiotechnol.*, *17*, 438–449.

Alshammari, S. O., Mahmoud, S. Y., & Farrag, E. S. (2023). Synthesis of green copper nanoparticles using medicinal plant Krameria sp. Root extract and its applications. *Molecules., 28*, 4629.

Álvarez, S. P., López, N. E. L., Lozano, J. M., Negrete, E. A. R., & Cervantes, M. E. S. (2016). Plant fungal disease management using nanobiotechnology as a tool. *Adv Appl Fungal Nanobiotechnol*, 169–192.

Ansari, M. A. (2023). Nanotechnology in food and plant science: Challenges and future prospects. *Plants. (Basel), 12*.

Aziz, N., Fatma, T., Varma, A., Prasad, R. (2014). Biogenic synthesis of silver nanoparticles using *Scenedesmus abundans* and evaluation of their antibacterial activity. *Journal of Nanoparticles*. http://dx.doi.org/10.1155/2014/689419

Aziz, N., Faraz, M., Pandey, R., Sakir, M., Fatma, T., Varma, A., Barman, I., Prasad, R. (2015). Facile algae-derived route to biogenic silver nanoparticles: Synthesis, antibacterial and photocatalytic properties. *Langmuir*. 31, 11605–11612. doi 10.1021/acs.langmuir.5b03081

Aziz, N., Pandey, R., Barman, I., Prasad, R. (2016). Leveraging the attributes of *Mucor hiemalis*-derived silver nanoparticles for a synergistic broad-spectrum antimicrobial platform. *Front Microbiol. 7*(1984). doi: 10.3389/fmicb.2016.01984

Aziz, N., Faraz, M., Sherwani, M. A., Fatma, T., Prasad, R. (2019). Illuminating the anticancerous efficacy of a new fungal chassis for silver nanoparticle synthesis. *Front Chem. 7*(65). doi: 10.3389/fchem.2019.00065

Caparco, A. A., Gonzalez-Gamboa, I., Hays, S. S., Pokorski, J. K., & Steinmetz, N. F. (2023). Delivery of nematicides using TMGMV-derived spherical nanoparticles. *Nano.Lett., 23,* 5785–5793.

Choudhary, N., Dhingra, N., Gacem, A., Yadav, V. K., Verma, R. K., Choudhary, M. et al. (2023). Towards further understanding the applications of endophytes: Enriched source of bioactive compounds and bio factories for nanoparticles. *Front Plant Sci., 14,* 1193573.

Daniel, A. I., Keyster, M., & Klein, A. (2023). Biogenic zinc oxide nanoparticles: A viable agricultural tool to control plant pathogenic fungi and its potential effects on soil and plants. *Sci.Total Environ., 897,* 165483.

de Moraes, A. C. P., Ribeiro, L. D. S., de Camargo, E. R., & Lacava, P. T. (2021). The potential of nanomaterials associated with plant growth-promoting bacteria in agriculture. 3 *Biotech*, *11*(7), 318.

Dhiman, S., Varma, A., Prasad, R., Goel, A. (2022). Mechanistic insight of the antifungal potential of green synthesized zinc oxide nanoparticles against *Alternaria brassicae. Journal of Nanomaterials*. 13 pages. https://doi.org/10.1155/2022/7138843

Dhiman, S., Varma, A., Rao, M., Prasad, R., Goel, A. (2023). Deciphering the fertilizing and disease suppression potential of phytofabricated zinc oxide nanoparticles on *Brassica juncea. Environmental Research*. https://doi.org/10.1016/j.envres.2023.116276

Gupta, N., Upadhyaya, C. P., Singh, A., Abd-Elsalam, K. A., Prasad, R. (2018). Applications of silver nanoparticles in plant protection. In: *Nanobiotechnology Applications in Plant Protection* (eds. Abd-Elsalam, K., and Prasad, R.), Springer International Publishing AG 247–266.

Gurpreet, K., & Singh, S. K. (2018). Review of nanoemulsion formulation and characterization techniques. *Indian J Pharm Sci*, *80*(5).

Hernández-Díaz, J. A., Garza-García, J. J., Zamudio-Ojeda, A., León-Morales, J. M., López-Velázquez, J. C., & García-Morales, S. (2021). Plant-mediated synthesis of nanoparticles and their antimicrobial activity against phytopathogens. *J Sci Food Agric*, *101*(4), 1270–1287.

Hussain, A., Yasar, M., Ahmad, G., Ijaz, M., Aziz, A., Nawaz, M. G. et al. (2023). Synthesis, characterization, and applications of iron oxide nanoparticles. *Int. J. Health Sci.(Qassim.)*, *17*, 3–10.

Ismail, M., Prasad, R., Ibrahim, A. I. M., Ahmed, I. S. A. (2017). Modern prospects of nanotechnology in plant pathology. In: *Nanotechnology* (eds. Prasad, R., Kumar, M., Kumar, V.), Springer Nature Singapore Pte Ltd. 305–317.

Kaushal, A., Khurana, I., Yadav, P., Allawadhi, P., Banothu, A. K., Neeradi, D. et al. (2023). Advances in therapeutic applications of silver nanoparticles. *Chem.Biol.Interact., 382*, 110590.

Kumari, R., Suman, K., Karmakar, S., Mishra, V., Lakra, S. G., Saurav, G. K. et al. (2023). Regulation and safety measures for nanotechnology-based agri-products. *Front Genome Ed., 5*, 1200987.

Linh, N. T., Qui, N. H., & Triatmojo, A. (2022). The effect of nano-encapsulated herbal essential oils on Poultry' s health. *Arch.Razi.Inst., 77*, 2013–2021.

Luque-Jacobo, C. M., Cespedes-Loayza, A. L., Echegaray-Ugarte, T. S., Cruz-Loayza, J. L., Cruz, I., de Carvalho, J. C. et al. (2023). Biogenic synthesis of copper nanoparticles: A systematic review of their features and main applications. *Molecules., 28*.

Malode, M. G. P., Chauhan, S. A., Bartare, S. A., Malode, L. M., Manwar, J. V., & Bakal, R. L. (2021). A critical review on nanoemulsion: Advantages, techniques and characterization. *J Appl Pharm Sci Res*, *4*(3), 6–12.

Mateo-Sagasta, J., Zadeh, S. M., & Turral, H. (Eds.). (2018). *More people, more food, worse water?: A global review of water pollution from agriculture*. FAO.

McClements, D. J., Das, A. K., Dhar, P., Nanda, P. K., & Chatterjee, N. (2021). Nanoemulsion-based technologies for delivering natural plant-based antimicrobials in foods. *Front Sustainable Food Syst*, *5*, 643208.

Mubeen, B., Hasnain, A., Wang, J., Zheng, H., Naqvi, S. A. H., Prasad, R., Rehman, A. U., Sohail, M. A., Hassan, M. Z., Farhan, M., Khan, M. A., Moustaf, M. (2023). Current progress and open challenges for combined toxic effects of manufactured nano-sized objects (MNO's) on soil biota and microbial community. *Coatings*. *13*, 212. https://doi.org/10.3390/coatings13010212

Mulqueen, P. (2003). Recent advances in agrochemical formulation. *Adv. Colloid Interface Sci*, 106(1–3), 83–107.

Nadaroglu, H., Güngör, A. A., & Selvi, İ. N. C. E. (2017). Synthesis of nanoparticles by green synthesis method. *Int. J. Innov. Res. Rev.*, *1*(1), 6–9.

Osherov, A., Prasad, R., Chrzanowski, W., New, E. J., Brazaca, L., Sadik, O., Haynes, C. L., Maine, E. (2023). Responsible nanotechnology for a sustainable future. *One Earth*. *6*(7), 763–766. https://doi.org/10.1016/j.oneear.2023.06.010

Prasad, R. (2014). Synthesis of silver nanoparticles in photosynthetic plants. *Journal of Nanoparticles*, 2014. http://dx.doi.org/10.1155/2014/963961

Prasad, R., Kumar, V., Prasad, K. S. (2014). Nanotechnology in sustainable agriculture: Present concerns and future aspects. *Afr J Biotechnol*. *13*(6), 705–713.

Prasad, R., Pandey, R., Barman, I. (2016). Engineering tailored nanoparticles with microbes: quo vadis. *WIREs Nanomed Nanobiotechnol 8*, 316–330. doi: 10.1002/wnan.1363

Prasad, R., Bhattacharyya, A., Nguyen, Q. D. (2017). Nanotechnology in sustainable agriculture: Recent developments, challenges, and perspectives. *Front Microbiol*. *8*, 1014. doi: 10.3389/fmicb.2017.01014

Peng, H., Jian, J., Long, H., Jiang, Q., Huang, W., Kong, L. et al. (2023). Self-Assembled nanonematicide induces adverse effects on oxidative stress, succinate dehydrogenase activity, and ATP generation. *ACS Appl.Mater.Interfaces.*, *15*, 31173–31184.

Sela, A., Shkuri, N., Tish, N., Vinokur, Y., Rodov, V., & Poverenov, E. (2023). Carboxymethyl chitosan-quercetin conjugate: A sustainable one-step synthesis and use for food preservation. *Carbohydr.Polym.*, *316*, 121084.

Shang, Y., Hasan, M. K., Ahammed, G. J., Li, M., Yin, H., & Zhou, J. (2019). Applications of nanotechnology in plant growth and crop protection: a review. *Molecules*, *24*(14), 2558.

Sharma, S., Loach, N., Gupta, S., & Mohan, L. (2020). Phyto-nanoemulsion: An emerging nano-insecticidal formulation. *Environ. Nanotechnol. Monit. Manag*, *14*, 100331.

Sindhu, M., Rajkumar, V., Annapoorani, C. A., Gunasekaran, C., & Kannan, M. (2023). Nanoencapsulation of garlic essential oil using chitosan nanopolymer and its antifungal and anti-aflatoxin B1 efficacy in vitro and in situ. *Int.J.Biol.Macromol.*, *243*, 125160.

Songca, S. P. (2023). Combinations of photodynamic therapy with other minimally invasive therapeutic technologies against cancer and microbial infections. *Int.J.Mol.Sci.*, *24*, 10875.

Srivastava, S., Bist, V., Srivastava, S., Singh, P. C., Trivedi, P. K., Asif, M. H., ... & Nautiyal, C. S. (2016). Unraveling aspects of Bacillus amyloliquefaciens mediated enhanced production of rice under biotic stress of Rhizoctonia solani. *Front Plant Sci*, *7*, 587.

Teodoro, K. B. R., Alvarenga, A. D., Rocha Oliveira, L. F., Marques Chagas, P. A., Lopes, R. G., Andre, R. D. S. et al. (2023). Fast fabrication of multifunctional PCL/Curcumin nanofibrous membranes for wound dressings. *ACS Appl.Bio Mater.*, *6*, 2325–2337.

Timmusk, S., Behers, L., Muthoni, J., Muraya, A., & Aronsson, A. C. (2017). Perspectives and challenges of microbial application for crop improvement. *Front Plant Sci*, *8*, 49.

Tiwari, J. N., Tiwari, R. N., & Kim, K. S. (2012). Zero-dimensional, one-dimensional, two-dimensional and three-dimensional nanostructured materials for advanced electrochemical energy devices. *Prog. Mater. Sci*, *57*(4), 724–803.

Yang, Y., Wang, J., Xiu, Z., & Alvarez, P. J. (2013). Impacts of silver nanoparticles on cellular and transcriptional activity of nitrogen-cycling bacteria. *Environ Toxicol Chem*, *32*(7), 1488–1494.

Yin, N., Zhang, Z., Ge, Y., Zhao, Y., Gu, Z., Yang, Y. et al. (2023). Polydopamine-based nanomedicines for efficient antiviral and secondary injury protection therapy. *Sci.Adv., 9*, eadf4098.

Zhang, H., Pan, Q., Cai, W., Shi, X., Yang, D. P., Lin, H. et al. (2023). C-doped ZnO nanocomposites molecularly imprinted photoelectrochemical sensor for ultrasensitive and selective detection of oxytetracycline in milk. *Food Chem., 426*, 136535.

Index

K

L

M

N

O

P

Q

R

S

T

V

W